CALCULATING INSTRUMENTS AND MACHINES

CALCULATING INSTRUMENTS AND MACHINES

DOUGLAS R. HARTREE

Plummer Professor of Mathematical Physics
University of Cambridge

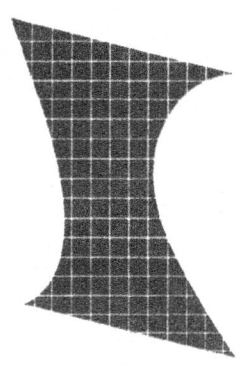

CAMBRIDGE
AT THE UNIVERSITY PRESS
1950

CAMBRIDGE UNIVERSITY PRESS
Cambridge, New York, Melbourne, Madrid, Cape Town,
Singapore, São Paulo, Delhi, Mexico City

Cambridge University Press
The Edinburgh Building, Cambridge CB2 8RU, UK

Published in the United States of America by Cambridge University Press, New York

www.cambridge.org
Information on this title: www.cambridge.org/9781107630659

First published 1950
First paperback edition 2012

A catalogue record for this publication is available from the British Library

ISBN 978-1-107-63065-9 Paperback

PREFACE

IN THE early fall of 1948 I visited the University of Illinois at the invitation of Dean Louis N. Ridenour, Dean of the Graduate School, to give a short series of lectures on calculating instruments and machines. These lectures, with only minor modifications, form the content of this book.

These lectures were to be devoted primarily to recent developments in the subject. But Dean Ridenour invited me to give some attention to its historical side, and I have been glad to do so if only to pay tribute to the remarkable vision and foresight, as it now seems, of two pioneers of thought in this field, Charles Babbage and Lord Kelvin.

Like the lectures, this book is intended as a general introduction to those who have no specialised knowledge in the subject, not as a detailed account for those already expert in it. To anyone who is, or has been, engaged in development work on any of the equipment mentioned, or on similar projects, this account will probably appear sketchy and inadequate; and probably no one group will consider that its own contribution to the subject is adequately represented. All I can hope is that this survey will be a useful introduction to the subject for those to whom it is primarily addressed.

In a series of lectures such as this, it seems appropriate for the lecturer to draw on his own first-hand experience, and to follow his own bias of interest, to a greater degree than would be suitable in a formal text-book, and this I have done, particularly in Chapter 3 and parts of Chapters 7, 8, and 9. If I am thought to have given too much prominence in Chapter 7 to two particular machines, the Harvard "Mark I Calculator" and the Eniac, I must explain that this prominence is deliberate; not because I happen to be better acquainted with these machines than with some others, but because I regard them as being outstanding steps in the development of automatic general-purpose machines, the one as the first practical realisation of such a machine and the other as the first electronic digital machine.

The subjects of calculating instruments (analogue machines) and calculating machines (digital machines) are here treated almost entirely separately, and the reader who is mainly interested in the latter can omit Chapters 2, 3, and 4 without missing anything important to the later argument. I have regarded desk machines and standard punched-card equipment as outside the scope of these lectures; the only digital machines with which I have been concerned are the automatic general-purpose machines.

In this field particularly, the subject is a live one, in which vigorous development is taking place. Since these lectures were given, a simple form of machine using Professor Williams' form of electrostatic storage (p. 96) has been put into operation at the University of Manchester (ref. 117).

v

Also quite recently the EDSAC at the Mathematical Laboratory of Cambridge University (p. 97) has been completed and proved to work satisfactorily, confirming that for this purpose a storage based on the use of mercury delay lines (p. 95) is a practicable project, and reports of satisfactory tests of a machine using the same form of storage have come from the United States. There is little doubt that the next few years will see further substantial developments in this field.

I wish to express my thanks to the Institute of Electrical Engineers for permission to draw on some of the material of my Kelvin Lecture to the Institution (ref. 90) for parts of Chapters 3 and 4; to the Metropolitan-Vickers Electrical Co. Ltd., to the Director of the Mathematical Laboratory, Cambridge University, and to Professor S. H. Caldwell for the photographs of differential analysers; to Professor H. H. Aiken for the photographs of the Harvard Mark I and Mark II Calculators; and to the U. S. War Department for the photographs of the Eniac.

Last, but not least, I am glad to take this opportunity to thank Dean Ricenour for the opportunity of giving these lectures at the University of Illinois, and to express my warm appreciation of the hospitality and friendly kindness I met there. I also wish to thank Dr. Wilbur Schramm and the staff of the University of Illinois Press for their cooperation in the production of these lectures in book form.

D. R. HARTREE

Cavendish Laboratory,
University of Cambridge,
England.

May, 1949

PREFACE TO THE ENGLISH EDITION

I WISH to thank the Cambridge University Press for undertaking the publication of this book in England, and the University of Illinois Press for agreeing to this course. I have taken the opportunity of this re-publication to make some corrections in the text and some additions to the list of references.

D. R. H.

May, 1950

CONTENTS

ix

Chapter 1

INTRODUCTION

It is convenient to distinguish two classes of equipment for carrying out numerical calculations by mechanical or electrical means. One class consists of those devices which operate by translating numbers into physical quantities of which the numbers are the *measures*, on specified scales, operating with these quantities, and finally *measuring* some physical quantity to get the required result. For example, a product xy may be evaluated by adjusting a variable resistance to have the value x ohms, then adjusting the current through it to have the value y milliamps., and measuring the potential difference across the resistance in millivolts (fig. 1). Other examples of devices of this class are the slide-rule, various

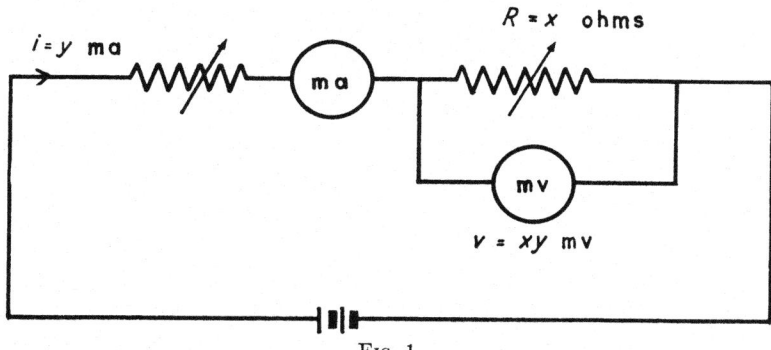

Fig. 1

forms of planimeter, integraph and harmonic analyser, the isograph, differential analyser, and cinema integraph.

The other class consists of those devices which take and operate with numbers directly in their digital form, usually, but not necessarily, by counting discrete objects such as the teeth of a gear wheel, or discrete events such as electrical pulses. Examples are the Brunsviga, Marchant, and other desk calculating machines, and standard I.B.M. equipment as used for computing purposes.

I have found it convenient to distinguish the two classes by the terms "instruments" and "machines" respectively; a corresponding distinction is made in the Encyclopaedia Britannica (14th Edition), in which the two classes of equipment are considered in different articles entitled "Mathematical Instruments" and "Calculating Machines" respectively. In America a similar distinction is made between "analogue machines" and "digital machines," but the former term is, I think, usually restricted

1

to the larger and more elaborate instruments; I have never heard a slide-rule referred to as an "analogue machine," though in my classification it is certainly an "instrument."

Of course, a single large piece of equipment may contain components of an "instrumental" character and others of a "machine" character, so that the whole may be of a composite character. But at the present stage of the art and science of the design and use of computing equipment, the distinction is a convenient one.

The two kinds of equipment have their characteristic advantages and limitations. Any one "instrument" is restricted to a rather narrow range of kinds of calculations; for example, a slide-rule to multiplication, division, and the evaluation of powers and roots; a differential analyser to the integration of ordinary differential equations. And further the accuracy of an instrument is restricted by the mechanical and electrical accuracy of its components and by the attainable accuracy of physical measurement of the result. On the other hand, it is possible to design instruments to deal with continuous variables, and in particular to carry out integration as a continuous process.

A "machine" can only handle numbers expressed in digital form to a finite number of significant figures; it cannot deal with continuous variables or continuous processes as such, and, in particular, in using a machine integration has to be replaced by summation over a finite number of finite intervals. On the other hand, a machine can be designed to work to any finite degree of accuracy without difficulty; in order to get, on a desk calculating machine such as a Marchant, a result accurate to 10 significant decimal figures, it is not necessary for any component to be constructed, or any measurement made, to an accuracy of 1 part in 10^{10}.

With an instrument, it is only possible to put an approximate question and to get an approximate answer. For example, with a slide-rule it is not possible to multiply exactly 2 by exactly 2.5; all one can do is to multiply a number in the range 2.000 ± 0.002, say, by one in the range 2.500 ± 0.002, and get a result perhaps in the range 5.000 ± 0.005. On the other hand, with a machine it is normally *only* possible to put an exact question and get an exact result (provided, of course, that the machine is operated and operating correctly). For example, we cannot multiply π by $\sqrt{2}$ on a machine; we can multiply 3.14159 by 1.41421, perhaps obtaining this product either exactly to ten decimal places or rounded off to, say, five places as required. The latter result might be regarded as an approximate answer to the question "find the product of 3.14159 by 1.41421," but from the point of view of how it is obtained on a machine, it is better regarded as an exact answer to another exact question, "find the integer part and first five decimals of $(3.14159 \times 1.41421 + 0.000005)$."

There have been considerable developments in both instruments and machines in the last twenty years. In the field of instruments the out-standing development has been the differential analyser, an instrument for evaluating by mechanical means the solution of ordinary differential equations. In the field of machines the main developments have been in the direction of large digital machines designed to carry out, rapidly and automatically, extended sequences of individual arithmetical operations.

The developments in these two fields have been almost independent, and their characters are so different that they can best be considered separately. The differential analyser will be the subject of Chapters 2 and 3, some other instruments will be considered in Chapter 4, and various machines will form the subject of Chapters 5 to 8.

Chapter 2

THE DIFFERENTIAL ANALYSER

2.1. The Nature of the Problem of Instrumental Solution of Differential Equations

Before considering the differential analyser, it will be as well to examine the general nature of the situation with which it deals, and what are the characteristic features of problems giving rise to this situation.

In many applications of mathematics to problems of pure and applied science, there occurs the idea of a *rate of change*, usually with respect to time or to space, of one or more of the relevant quantities, and then the idea of integration is involved in the process of finding the *total* change in a time or space interval from the *rate* of change, which will usually be varying through the interval. In contexts of this nature there are two different kinds of situation which may arise, of which simple examples are provided by the responses of two different circuits to an applied voltage, varying in a known way with time (fig. 2).

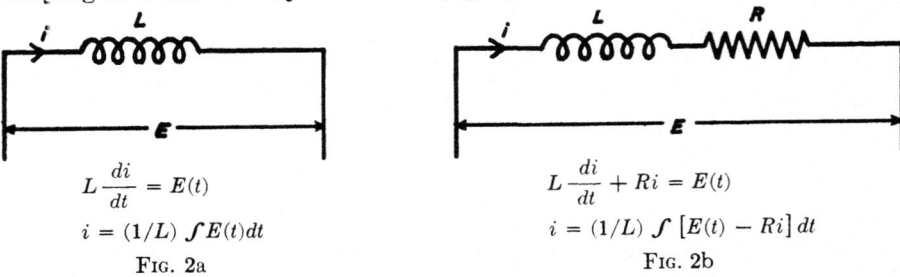

$$L \frac{di}{dt} = E(t)$$

$$i = (1/L) \int E(t)dt$$

Fig. 2a

$$L \frac{di}{dt} + Ri = E(t)$$

$$i = (1/L) \int [E(t) - Ri] \, dt$$

Fig. 2b

If the voltage is applied to a circuit with inductance but with negligible resistance (as in fig. 2a), the time rate of change of current at any moment depends only on the value of the voltage at that moment, which is given and is independent of the value the current itself happens to have. Then we are concerned simply with the evaluation of an integral whose integrand is a known function of the independent variable.

But it very often happens that the rate of change of a quantity at any moment (or point) depends on the magnitude of that quantity itself at that moment (or point). For example, if the voltage is applied to a circuit with inductance and resistance (as in fig. 2b), the rate of change of current depends on the instantaneous value of the current itself, as well as on the voltage. The formal expression of such a situation is what is called in mathematics a *differential equation*, and the determination of the current at any time involves the solution of this differential equation. This solution can be regarded as the result of evaluating an integral in which the integrand at any time depends in a definite way on the result of the integration up to that time. For example in the circuit illustrated in fig. 2b, the equation for the current can be written

$$i = (1/L) \int [E(t) - Ri]\, dt,$$

and in the integral here, the current i to be found occurs in one contribution to the integrand.

This aspect of a differential equation is not prominent in the conventional formal treatment of such equations, but it expresses rather closely the way in which it is often best to consider their mechanical solution. Indeed, from the point of view of mechanical integration, it is just this feature which distinguishes the evaluation of a solution of a differential equation from the evaluation of a simple integral of a known function of the independent variable. Thus the essential points in a mechanical instrument for integrating differential equations are an integrating mechanism and means of furnishing to that mechanism as integrand a quantity depending in a definite way on the value of the integral calculated by it.

2.2. Integrating Mechanisms

Any continuously variable gear can be used as an integrating mechanism. For suppose that in fig. 3 the rectangle represents any mechanism giving a continuously variable gear ratio $1:n$ between the rotations of driving and driven shafts. Then for a rotation dx turns of the driving shaft, at gear ratio $1:n$, the rotation of the driven shaft is $n\,dx$ turns. If the gear ratio n is changing as the driving shaft is rotating, the total rotation of the driven shaft is the sum of the elements of rotation $n\,dx$; that is, it is $\int n\,dx$ turns.

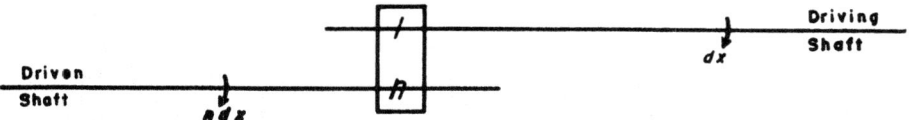

FIG. 3. Continuously variable gear as integrator.

To be suitable for incorporation in a differential analyser, such a mechanism should be able to be set accurately to any gear ratio n in its range and should include both positive and negative values of n, including $n=0$, in this range. The form used in most differential analysers consists of a vertical wheel driven by a horizontal disc (fig. 4),

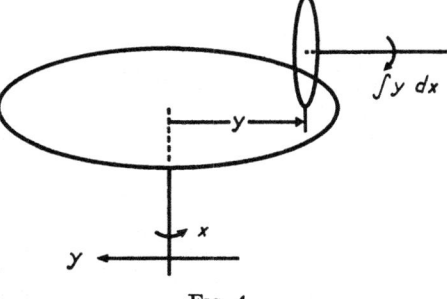

FIG. 4

the wheel and disc being so mounted that the distance between the centre of rotation of the disc and the point of contact of the wheel with it can be varied. The gear ratio between the rotations of disc and wheel is proportional to this distance, which is usually called the "displacement" of the integrator; this must therefore be made proportional to the integrand of the integral to be evaluated. The rotation of the disc, usually called the "rotation" of the integrator, represents the variable of integration and must be made proportional to it. It is interesting that a planimeter based on this integrating mechanism is considerably older than the now more familiar Amsler planimeter (see ref. 31).

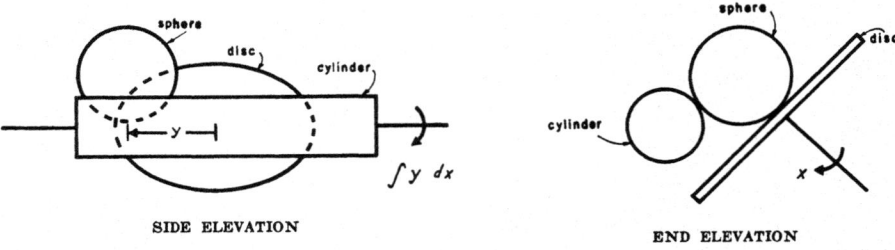

SIDE ELEVATION   END ELEVATION

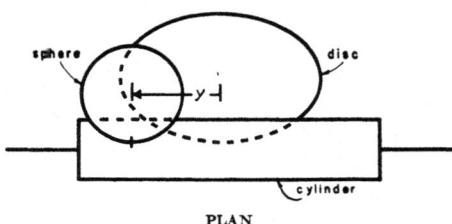

PLAN

FIG. 5. James Thomson's integrator.

Another form of integrator (fig. 5), devised by James Thomson, brother of Lord Kelvin, consists of an inclined disc, a horizontal cylinder not in contact with it, and a sphere in contact with both (104). The sphere can be moved along the cylinder, and makes contact with the disc along its horizontal diameter. It gives a drive from disc to cylinder at a gear ratio which varies proportionately to the distance from the centre of the disc to the point at which the sphere is in contact with it.

2.3. The General Idea of the Differential Analyser

The differential analyser consists essentially of a number of integrating mechanisms which can be connected together so as to evaluate solutions of ordinary differential equations.

The idea of connecting together integrating mechanisms for this purpose is not by any means new. It was originally formulated, clearly and in considerable detail, by Kelvin in two papers published as long ago as 1876 (106, 107). But the realisation of this idea in the form of a practical working instrument was an achievement of Dr. Bush and an able group working with him at M.I.T. nearly twenty years ago (17) (see also refs.

27, 47, 49, 58); and it should be added that their conception of such a machine was independent of Kelvin's.

It is interesting to follow in Kelvin's papers the development of his thought on the subject. His brother, James Thomson, had recently designed the integrating mechanism shown in fig. 5, and Kelvin had seen how it could be adapted to evaluate continuously the integral of the product of two functions, and had outlined its application to a harmonic analyser (105).

His first paper on its application to the integration of differential equations is concerned with the equation

$$\frac{d}{dx}\left[\frac{1}{F(x)}\frac{dy}{dx}\right] + y = 0,$$

of which he says: "On account of the great importance of this equation in mathematical physics" (of which he gives examples) "I have long endeavoured to obtain a means of facilitating its practical solution. . . ." After some general remarks he continues, "A ready means of obtaining approximate results which shall show the general character of the solutions . . . has always seemed to me a desideratum. Therefore I have made many attempts to plan a mechanical integrator which should give solutions by successive approximations." He then shows how an instrument consisting of two of James Thomson's integrators, connected together so that the result of the integration performed by the first forms the integrand for the second, will provide such a result by an iterative process, as follows. If an integrand y_0 is fed to the first of the two integrators of such an instrument, the output from the second is

$$y_1 = \int F(x)\left[c - \int y_0\, dx\right] dx;$$

if this is recorded, and subsequently fed as integrand to the first integrator, the output from the second is

$$y_2 = \int F(x)\left[c - \int y_1\, dx\right] dx$$

and so on. The process is continued until there is no appreciable difference between input and output, and the common input and output is then the solution of the equation.

Then comes the inspiration, and here again it is worth quoting Kelvin's own words. "So far I had gone and was feeling satisfied, feeling I had done what I wished to do for many years. But then came a pleasing surprise. Compel agreement between the function fed into the double machine and that given out by it." He then shows how, in principle, this can be done by making a second interconnection between the two integrators so that the output of the second is used continuously as the integrand of the first, and that this interconnected system of integrators evaluates the solution of the equation directly. He continues: "Thus I was led to a conclusion which was wholly unexpected; and it seems to me very remarkable that the general differential equation of the second order with variable coefficients may be rigorously, continuously, and in a single process solved by a machine."

It is clear from the context that here he has in mind only linear second-order equations. But in a second paper he is concerned with the extension to linear equations of any order, and in an addendum includes in principle the further extension to non-linear equations, even including the idea of three-dimensional cams for feeding in functions of two independent variables.

Harmonic analysers using James Thomson's integrating mechanism as suggested by Kelvin have been constructed (108). But as far as I have been able to ascertain, no steps were taken at the time for realising his conception of an instrument for solving differential equations, and, as already mentioned, it was left to Dr. Bush and his team at M.I.T. to develop in the differential analyser a machine which could be built and which would be accurate and reliable in operation.

2.4. General Structure of the Earlier Forms of Differential Analyser

A differential analyser consists of a number of units, each of which carries out an operation which can be regarded as a translation into mechanical terms of a process (integration, addition, etc.) which may be required in the mechanical integration of a differential equation, and some means of interconnecting these units. Each unit is driven by the rotation of one or more shafts, and the result of its operation is the rotation of a shaft driven by it. Each shaft represents one of the quantities occurring in the equation to be solved, and the total rotation of each shaft measures the corresponding quantity, on an assigned scale. The interconnections between the units are made in such a way that the relations between the rotations of the various shafts form a translation into mechanical terms of the equation to be solved. Then the rotation of one shaft, representing the independent variable, drives the remainder of the shafts in accordance with the equation represented by these interconnections.

In the original differential analyser, and in others which have been built on the same general plan though with various differences of detailed design, the connections between the units are purely mechanical, by means of shafts and gearing. A later development will be considered in §2.5.

The instrument with which I personally have been most closely associated is that installed at the University of Manchester, England, in 1935 (87); this is of the earlier type with purely mechanical interconnections, for which it will serve as an example. A similar instrument, at the

Mathematical Laboratory of the University of Cambridge, is illustrated in fig. 6.

The main units, the integrators, are arranged in pairs in cases along the near side. Means must be provided for supplying the instrument with information about functional relationships between variables occurring in

FIG. 6. Differential Analyser at Mathematical Laboratory, University of Cambridge.

the equation, as for example the relation between voltage and current in a non-linear resistive element in a circuit. In some cases this can be done by generating the required relation by the solution of an auxiliary equation. In others it is necessary to supply the information from an "input table"; four such tables are on the far side of the main frame in fig. 6. Means must also be provided for recording the result; this can be done either in graphical form on an "output table" or in numerical form on a set of counters.

The shafts directly driving or driven by the various units lie across the length of the main frame, and are connected through cross-drives and longitudinal shafts. Adding units can be incorporated in the various gear trains in these interconnections.

Some close-up views of individual components are given in figs. 7, 8, 10. Figure 7 shows two integrators with associated equipment. Each integrator is a precision form of continuously variable gear of the disc-and-wheel type already mentioned (fig. 4). In the interconnections between the various units required in the solution of a differential equation, the output from an integrator may have to drive several other units, and this may form much too heavy a load to be driven directly by the friction between the disc and wheel of an integrator. Hence on the output side

FIG. 7. Integrator and torque amplifier.

there is a torque amplifier, which drives the output shaft at the same speed as the integrating wheel, but with a much greater torque. In the instrument illustrated in fig. 6, as in Bush's original differential analyser, the torque amplifier is a purely mechanical servo operating on the capstan principle. In some other more recent instruments this has been replaced by an electrical servo, for example by an amplidyne system controlled by an optical signal measuring the angular error between the positions of the integrating wheel and of the output shaft.

Figure 8 shows an input table. A bridge spans the table and can be moved perpendicularly to its length by the input shaft to the table. The bridge supports a carriage which can be moved along its length by the rotation of a handle, and which carries an index. A graph, representing the information to be fed to the differential analyser, is placed on the table. As the bridge is traversed across the table by the rotation of the input shaft, an operator turns the handle so as to keep the index on the curve; the human operator may be replaced by an automatic curve-follower (11, 65). A drive from the handle is taken to one of the cross shafts of the machine which forms the output shaft of the unit, from which the information expressed by the curve can be taken to the input shaft of the unit where it can be used (for example, to the displacement of an integrator).

The output table is similar, but carries a pencil or pen whose motions in both co-ordinates are driven by the rotation of input shafts, so that it draws a graph of the solution of the equation.

A unit incorporated in both the differential analysers built in England, but not, as far as I know, in any other, is a special input-output table by means of which an index can be made to follow, in the course of a solution, a curve drawn earlier by a pencil in the course of the same solution. This unit is similar to an input table, but is spanned by two parallel bridges which can be moved independently. Each bridge supports a carriage which can move along its length; one carriage carries a pen and the other an index. By means of this unit it is possible to evaluate solutions of equations in which the rate of change of a quantity, say x, at time t depends not only on what the value of x is at time t, but also on the value of x at a previous time $t - \tau$, where the "time-lag" τ may be constant or variable.

FIG. 8. Input table.

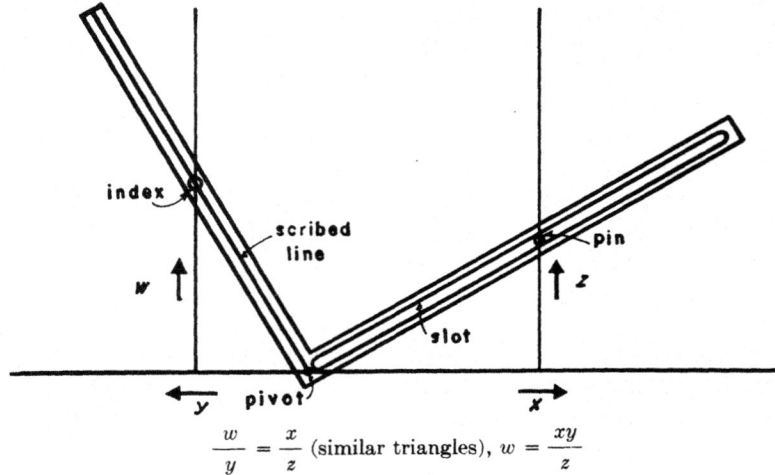

$$\frac{w}{y} = \frac{x}{z} \text{ (similar triangles)}, \ w = \frac{xy}{z}$$

FIG. 9. Multiplier-divider unit.

This unit has found another use, for which it was not originally planned. Its three independent inputs and one output make it very convenient for use as a multiplier-divider unit using the principle of similar triangles as in the multiplier of Bush's differential analyser. In a piece of sheet steel of L-form, a slot is cut in one leg of the L and a line is scribed at right angles to the slot on the other leg; the whole is pivoted at the intersection of the line and the axis of the slot (fig. 9). The pen of the special table is replaced by a rod which is a close fit in the slot; then if the index is moved

FIG. 10. Interconnections.

FIG. 11. Model version of differential analyser.

so as to lie on the line as the three inputs x, y, z vary, the ordinate of the index is xy/z.

Figure 10 shows some details of the interconnections between the various units, helical gears for the connections between the longitudinal and cross shafts, adding units, and a "frontlash" unit for compensating backlash in the gear train between the output shaft of one unit and an input shaft of another. The longitudinal shafts are on a different level from the cross shafts, so that the cross shaft driven by one unit can be connected to any one of the longitudinal shafts, and this again to any cross shaft driving any unit. This freedom of interconnection is very important, and one result of it is the very wide range of equations to which the differential analyser can be applied. The whole set of interconnections required for the solution of a given equation is called shortly the "set-up" for that equation.

Another important feature is that the relations between the rotations of the different shafts of the analyser are purely kinematical, and are independent of the actual speeds of rotation. Hence the operation of the analyser is independent of the speed of the shaft whose rotation represents the independent variable, and it is not even necessary that this speed should be constant.

In its accurate form, the differential analyser is a large and costly instrument. Various smaller and rougher versions of it have been built (8, 14, 59, 79), the first by Dr. Porter and myself. This instrument (fig. 11) was constructed largely of standard Meccano parts, and was successful beyond our expectations. It gave an accuracy of the order of 2%, and could be, and was, used for serious quantitative work for which this accuracy was adequate.

2.5. A New Differential Analyser

In a later form of differential analyser developed at M.I.T. by Bush and Caldwell (18), the interconnections, instead of being purely mechanical as in the instrument illustrated in fig. 6, are electrical. The individual units are still mechanical, and in principle the same as in the earlier form, though different in mechanical design and in appearance; but the assembly of cross and longitudinal shafts and cross-drives for interconnecting them is replaced by an electrical system. The output shaft of each unit carries an electrical angle-indicator consisting of a pair of variable condensers of which the capacities are $C_0 + C_1 \cos \phi$ and $C_0 + C_1 \sin \phi$ where ϕ is the azimuth of the shaft about its axis. Each input shaft carries a similar angle-indicator. To drive an input shaft A of one unit from the output shaft B of another (or the same) unit, the angle-indicators on the two shafts are connected so that they give an electrical signal which is a measure of the difference in the azimuths ϕ_A and ϕ_B of the two shafts. A motor driving the shaft A is controlled by this signal (and its time derivative and integral) so as to keep this difference $\phi_A - \phi_B$ to a small fraction of a turn under all operating conditions of angular velocity and acceleration. A very schematic diagram of the circuit is given in figs. 12 and 13.

Interconnections are made at a low power level, and standard automatic telephone switching equipment is used for setting them up. The interconnections required are punched in coded form on a paper tape. This is fed through a reading unit, in which electrical circuits are made through the punched holes. Currents in these circuits then energise relays which make the required connections.

This differential analyser has a number of electrical and mechanical

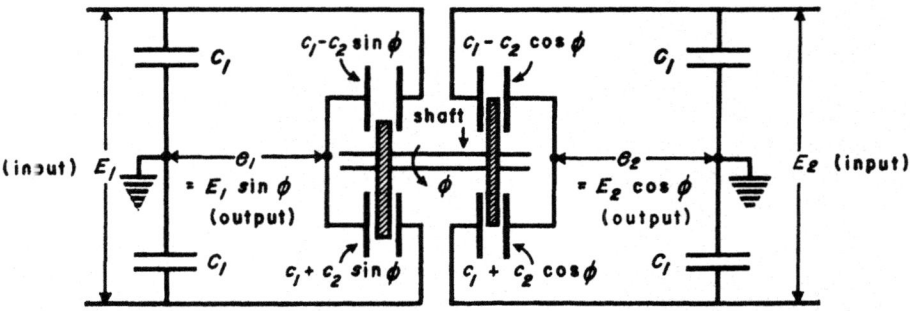

Fig. 12. Capacity bridge circuits of electrostatic angle-indicator.

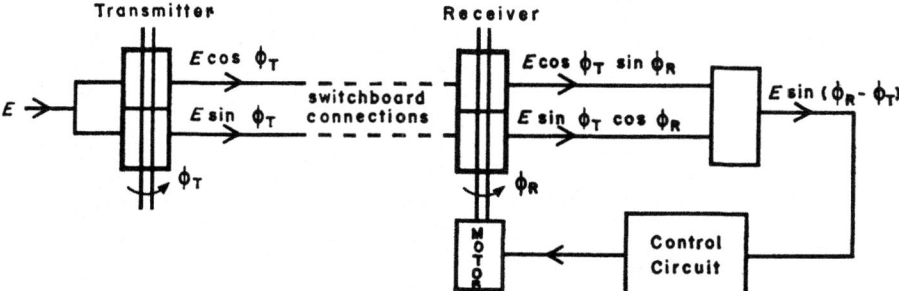

FIG. 13. Interconnections in the new differential analyser.

refinements such as multi-ratio gear boxes enabling any ratio expressed by a four-digit decimal fraction to be set up automatically from a second punched tape, means for setting initial displacements of integrators automatically from a third punched tape, and electric typewriters for automatically recording results in numerical form, as well as an output table for recording them graphically. It has 18 integrators, with scope for expansion to 30, and means for running simultaneously three independent problems if within the capacity of the available units.

Three separate punched tapes are used to supply to the instrument the information required in any one solution, the information on the three tapes being of different degrees of generality. That on one, the "A-tape," specifies the required interconnections of the various units and is an expression of the general structure of the equation, without reference to numerical values of coefficients or of the forms of functions supplied from input tables. That on another, the "B-tape," specifies values of ratios in any gearboxes used, which are determined by the numerical values of coefficients and by scale factors; this is common to all solutions, with different initial conditions, of the same equation. The third, the "C-tape," specifies initial displacements of integrators, which will usually be different from one solution to another.

Figure 14 shows a general view of the operating and control room; in the foreground are three input tables, of cylindrical form; on the left are the reading units for the punched tapes; to the right of them are the output table and control panel and to the right of this two of the electrical typewriters; in the background there are monitor panels and other equipment. Figure 15 shows an integrator and fig. 16 a portion of a typical equipment bay.

One of the important features of this form of differential analyser is the ease of changing over from the interconnections required for one equation to those for another. On a differential analyser of the earlier

FIG. 14. Operating room.

FIG. 15. Integrator of the new differential analyser.

type, such as that illustrated in fig. 6, such a change of set-up involves changing by hand all the cross-drives, gear wheels, adding units and couplings involved in making the mechanical interconnections. This takes some time, perhaps a day or two including the checking of the new set-up. On a differential analyser of the newer type all that is required is to replace the A- and B-tapes for one equation by those for the other. This, and the automatic process of setting up the connections and gear ratios from the punched tapes, takes only a few minutes.

This type of interconnection is also more flexible. For example, in the normal use of an integrator, the rotation of the integrating disc x is an "input" and that of

Fig. 16. Equipment bay.

the wheel z is an "output." But it is possible to modify the connections so that the motor driving the disc does so in such a way that the rotation of the wheel agrees with the rotation of the output shaft of some other unit; that is, the shaft representing x is driven so that $\int y dx$ is made equal to a given rotation z, or $x = \int dz/y$. This, which is equivalent to driving the disc from the wheel, cannot be done with an integrator fitted with a mechanical torque amplifier.

Another advantage is the greater flexibility of geographical arrangement of the components of the machine. With purely mechanical interconnections as in the example illustrated in fig. 6, questions of access and of mechanical design of the system of interconnections restrict rather severely the locations of the separate units; with electrical interconnections they can be placed wherever most convenient.

2.6. Using the Differential Analyser

Though the differential analysers illustrated in figs. 6 and 14 are so different in construction and appearance, the general way in which one has to think of a differential equation, from the point of view of planning a machine set-up to evaluate its solutions, is the same for both instruments, and the same notation can be used to represent the system of interconnections required. This notation can be thought of as representing, in schematic form, a plan of an instrument with mechanical connections, as set up to solve the equation considered.

The different units are represented by symbols shown in fig. 17, in which the relations between the quantities represented by the rotations of the driving and driven shafts of each kind of unit are indicated. In a set-up diagram, fig. 18 for example, they are shown connected by lines

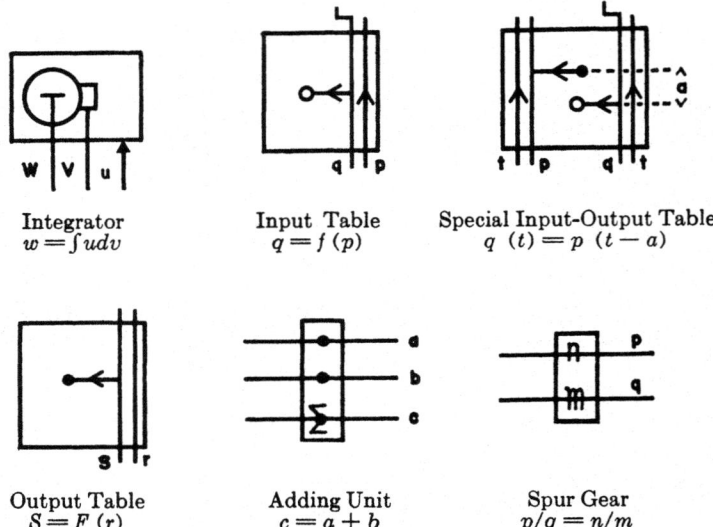

| Integrator | Input Table | Special Input-Output Table |
| $w = \int u \, dv$ | $q = f(p)$ | $q(t) = p(t - a)$ |

| Output Table | Adding Unit | Spur Gear |
| $S = F(r)$ | $c = a + b$ | $p/q = n/m$ |

Fig. 17. Notation for units of differential analyser in set-up diagrams.

representative of the cross shafts and longitudinal shafts, and each line is labelled with the quantity represented by the rotation of the corresponding shaft.

In the great majority of equations which arise in practice, the highest derivative of the dependent variable y is expressed in terms of the independent variable x and the lower derivatives of y, as for example in

$$\frac{d^2y}{dx^2} = y \cos(x + y) - \left(\frac{dy}{dx}\right)^2. \tag{2.1}$$

Then in planning a set-up, it is convenient to consider the equation in the form it takes when integrated once with respect to the independent variable, for example in this case

$$\frac{dy}{dx} = \int y \cos(x + y) \, dx - \int \left(\frac{dy}{dx}\right) dy. \tag{2.2}$$

This form shows that it is not necessary explicitly to construct the quantity $(dy/dx)^2$ in (2.1). A set-up for this equation is shown in fig. 18; the quantity $\cos(x + y)$ is generated by integrators V and VI by solution of the auxiliary equation $d^2w/dz^2 = -w$ with $z = x + y$, and the first of the integrals in equation (2.2) is evaluated by integrators III and IV, in the form $\int \cos(x + y)\, d\{\int y\,dx\}$; this is a standard connection for the integral of a product. The second of the integrals in (2.2) is evaluated by integrator II, and $y = \int (dy/dx)\, dx$ is evaluated by integrator I.

At a later stage of planning the set-up, scale factors (for example the number of turns of the x and y shafts which are to represent one unit of x and one unit of y respectively) are introduced and arrangements are made for obtaining the required signs in the relations between the rotations of the various shafts; but this aspect will not be considered in detail here. (For details, see references 17, 27, 47.)

This set-up provides an example of the use of a part of the machine to generate a function occurring in the main equation, in this case the function $\cos(x + y)$, by the solution of an auxiliary differential equation. This is a common feature of the use of the differential analyser, and can be used, as in this example, to generate functions of any variable, not

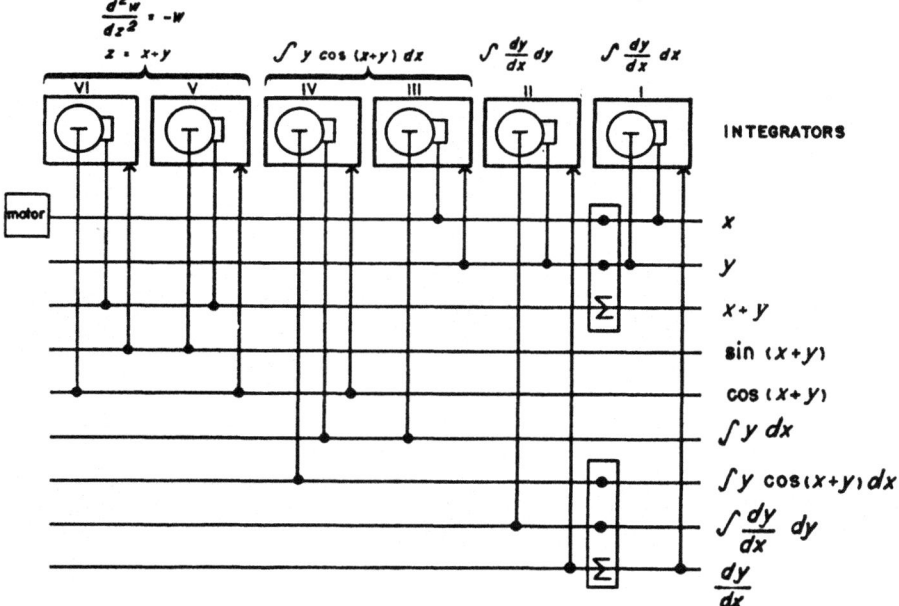

FIG. 18. Schematic set-up of differential analyser for equation:
$$\frac{d^2y}{dx^2} - y\cos(x + y) + \left(\frac{dy}{dx}\right)^2 = 0, \text{ or } \frac{dy}{dx} = \int y\cos(x + y)dx - \int \frac{dy}{dx}\, dy.$$
(Scale factors and signs omitted.)

only of the independent variable. Other simple functions of any variable w can be generated in a similar way, for example $1/w$, w^2, $w^{\frac{1}{2}}$, $\log w$, $\tan^{-1}w$, $e^{\beta w}$ (β being a positive or negative constant).

2.7. Regenerative Connections

An interesting group of connections, sometimes called "regenerative" connections, is obtained by feeding back the output from an integrator as a contribution to its own rotation. In the connection shown in fig. 19, the integrator displacement z and the contribution w to the integrator rotation are to be regarded as inputs; then

$$dy = z\,d(w + y) \tag{2.3}$$

so that the outputs are

$$y = \int \frac{z}{1 - z}\,dw, \qquad w + y = \int \frac{1}{1 - z}\,dw. \tag{2.4}$$

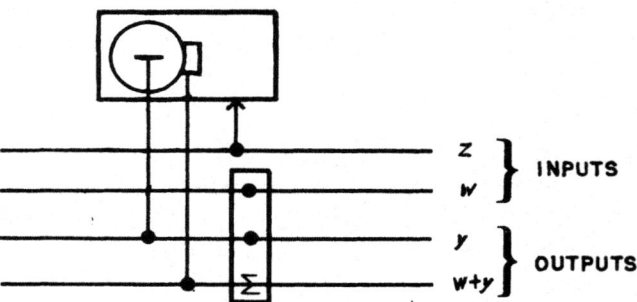

Fig. 19. Regenerative connection (Amble).

If z is given a constant value between 0 and 1, use of the output $w+y$ gives a means of gearing up a rotation without loss of torque (the additional power comes from the motor driving the torque amplifier). If $(1-z)$ is made equal to w, it gives $(w+y)=\log w$, providing a means of generating $\log w$ with one integrator; without regenerative connections two integrators are required. And if $(1-z)$ is made equal to $(w+y)$, (2.3) gives $(w + y) = (2w)^{\frac{1}{2}}$, giving a means of generating $w^{\frac{1}{2}}$ with one integrator.

Other regenerative connections involving two integrators can be used for obtaining z/y, $\int (z/y)dx$, $w^{m/n}$ where m and n are integers. A systematic study of such connections has been made by Amble (5); a few of the simpler ones involving one integrator, such as (2.4) with z constant, had been known and used previously, but Amble's work was independent. A further study of these and similar connections has been made by Michel (84).

Another example, in which a regenerative connection was used to avoid an inconvenient division, occurred in connection with the equation (122)

$$y'' + (A + e^{-x}/x)y = 0. \tag{2.5}$$

This can be written

$$ay' = -\smallint(e^{-x} + Ax)y\,dx - \smallint(x - a)\,dy'$$

where a is any convenient constant. With the equation in this form, e^{-x} can be generated in the standard way, and the awkward division by x, which would appear to be required from the equation in the form (2.5), is avoided. The occurrence of an integral with respect to y' in the expression for y' is an indication that a regenerative connection is required. This method of dealing with this situation was devised by Miss E. Monroe, while working with the differential analyser at the Mathematical Laboratory at the University of Cambridge; as far as I am aware, it is the first use of a regenerative connection.

2.8. Automatic Generation of Any Function

Reference has already been made to the generation of functions of certain definite analytical form, such as w^2, e^w, $\cos w$, $\log w$, by the solution of auxiliary differential equations. A further step is to generate automatically any function, and so avoid the use of input tables altogether.

This can be done by using part of the differential analyser, or a special unit, to act as an interpolating mechanism, so that if supplied with any function $f(w)$ at a set of tabular values of w, it will interpolate $f(w)$ as a continuous function of w for intermediate values of w. There are various ways of doing this; the following method is based on one shown to me by Dr. Perry Crawford, then of the Centre of Analysis, M.I.T. It will be supposed that $f(w)$ is tabulated at equal intervals of w; the "central difference" notation will be used for the finite differences of $f(w)$ (see fig. 20).

x	y				
x_{-1}	y_{-1}		$\delta^2 y_{-1}$		$\delta^4 y_{-1}$
		$\delta y_{-\frac{1}{2}}$		$\delta^3 y_{-\frac{1}{2}}$	
x_0	y_0		$\delta^2 y_0$		$\delta^4 y_0$
		$\delta y_{\frac{1}{2}}$		$\delta^3 y_{\frac{1}{2}}$	
x_1	y_1		$\delta^2 y_1$		$\delta^4 y_1$ etc.
		$\delta y_{1\frac{1}{2}}$		$\delta^3 y_{1\frac{1}{2}}$	
x_2	y_2		$\delta^2 y_2$		$\delta^4 y_2$
		$\delta y_{2\frac{1}{2}}$		$\delta^3 y_{2\frac{1}{2}}$	
x_3	y_3		$\delta^2 y_3$		$\delta^4 y_3$

etc.

in general

$$\delta y_{n+\frac{1}{2}} = y_{n+1} - y_n; \quad \delta^2 y_n = \delta y_{n+\frac{1}{2}} - \delta y_{n-\frac{1}{2}};$$
$$\delta^k y_n = \delta^{k-1} y_{n+\frac{1}{2}} - \delta^{k-1} y_{n-\frac{1}{2}}$$

FIG. 20. Central difference notation for finite differences.
y tabulated at equal intervals δx of x.

If an interpolation formula correct to third differences is used as the basis of the interpolation, then between any two tabular values, $f(w)$ is represented by a cubic in w, and such a function can be generated by a set of three integrators.

For interpolation in the interval $w = w_0$ to $w = w_1 = w_0 + \delta w$, the Newton-Gauss interpolation formula, correct to third differences, is

$$f(w_0 + \theta\delta w) = f_0 + \theta\delta f_{\frac{1}{2}} + \tfrac{1}{2}\theta(\theta - 1)\delta^2 f_0 + \tfrac{1}{6}\theta(\theta + 1)(\theta - 1)\delta^3 f_{\frac{1}{2}}, \quad (2.6)$$

and for interpolation in the interval $w = w_{-1} = w_0 - \delta w$ to w_0, the corresponding formula is

$$f(w_{-1} + \theta\delta w) = f_{-1} + \theta\delta f_{-\frac{1}{2}} + \tfrac{1}{2}\theta(\theta - 1)\delta^2 f_{-1} + \tfrac{1}{6}\theta(\theta + 1)(\theta - 1)\delta^3 f_{-\frac{1}{2}}. \quad (2.7)$$

If we differentiate formula (2.7) and put $\theta = 1$, we get

$$(\delta w)f_0' = \delta f_{-\frac{1}{2}} + \tfrac{1}{2}\delta^2 f_{-1} + \tfrac{1}{3}\delta^3 f_{-\frac{1}{2}}, \quad (2.8)$$

whereas if we differentiate (2.6) and put $\theta = 0$, we get

$$(\delta w)f_0' = \delta f_{\frac{1}{2}} - \tfrac{1}{2}\delta^2 f_0 - \tfrac{1}{6}\delta^3 f_{\frac{1}{2}}. \quad (2.9)$$

The difference between expressions (2.8) and (2.9) is $-\tfrac{1}{6}\delta^4 f_0$. Thus as w increases through w_0, a change $\Delta f_0'$ given by

$$(\delta w)\,\Delta f_0' = -\tfrac{1}{6}\delta^4 f_0 \quad (2.10)$$

has to be made in f' in order to change from the interpolation formula appropriate to the interval w_{-1} to w_0 to that for the interval w_0 to w_1. The corresponding changes in f'' and f''' are

$$(\delta w)^2\,\Delta f_0'' = 0 \quad (2.11)$$

$$(\delta w)^3\,\Delta f_0''' = \delta^4 f_0. \quad (2.12)$$

Thus three integrators, with provision for changing the displacements of two of them by the amounts (2.10) and (2.12) as w passes through each tabular value, will generate the function $f(w)$ continuously to the accuracy of the interpolation formula (2.6).

In terms of the last figure retained in f as unit, the quantities f, $6(\delta w)f'$, $2(\delta w)^2 f''$ and $(\delta w)^3 f'''$ must all be integral at each tabular value of w, and this can be used to provide a control on the accuracy with which $f(w)$ is being generated, and to correct minor errors.

This application of a set of integrators to form an instrument for continuous interpolation can be used to draw input curves from functions specified only by tables of values. The differential analyser itself has been used to draw the curves used in one form of automatic curve follower on the differential analyser at M.I.T., and a continuous interpolator has been built as a separate instrument by the Reeves Instrument Corporation.

2.9. Other Forms of Differential Analyser

The "differential analyser" was the name originally given to the mechanical instrument for integrating differential equations, developed by Bush and his group at M.I.T. But other instruments, considerably different in principle, can be designed for the same purpose, and it is convenient to use the term as a generic name for such instruments. Two will be mentioned shortly here.

One uses an electromagnetic method for carrying out integration, by an application of a component, developed for radar purposes during the war, called a "velodyne" (119). A velodyne consists (fig. 21) of a split-field motor and tacho-generator (a generator designed to be used as a tachometer) on a common shaft, the field current to the motor being supplied by a control circuit which takes the difference between an applied voltage V and the tacho-generator output, and drives the motor so as to keep this difference within a small range of tolerance. Thus if θ is the angle turned through by the motor,

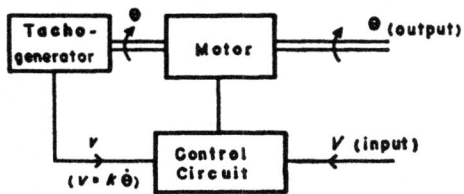

FIG. 21. Principle of the velodyne.

$$\frac{d\theta}{dt} = kV, \qquad \theta = k \int V dt \qquad (2.13)$$

where k is a constant. Thus the velodyne, regarded as an integrator, takes an electrical signal for the integrand and gives out a mechanical rotation to represent the integral. This rotation can then be transformed into an electrical voltage signal by means of a potential divider. Multiplication can be carried out by means of potential dividers, with electronic buffer circuits so arranged that no current, only a voltage, is taken off by the variable tapping on the potential divider.

It will be noted from (2.13) that integration is always with respect to actual time t. In this feature this instrument differs considerably from the mechanical differential analyser in which, as already mentioned, the rate of running of the machine in actual time is (within limits) irrelevant to its operation. Normally t is taken as a measure of the independent variable, and integration of a quantity z with respect to any other variable of the problem, say y, has to be carried out in the form $\int zy'dt$. However, since a voltage proportional to the product of two of the variables of the problem can easily be produced by potential-divider circuits, the evaluation of such integrals offers no difficulty. A four-integrator instrument of this type has been built in England by Dr. Uttley.

Another type of differential analyser is one using purely electrical methods for carrying out the various operations required. Feedback amplifers can be used to obtain integrating and differentiating circuits, the independent variable being actual time t as in the velodyne integrator; and multiplication can be carried out by potential-divider circuits as in Uttley's electromechanical differential analyser.

An instrument of this kind, using d.c. amplifiers, has recently been put on the market by the Reeves Instrument Corporation (94). In this instrument, the variables in the equation are represented by time-varying voltages. Addition, multiplication by constants, and integration are carried out by electrical circuits. Multiplication of two variables, each represented by a voltage, is carried out by applying one voltage to a potential divider, and using an electromechanical servo to locate the moveable contact at a position which depends linearly on the other voltage. Similar servos, operating non-linear potentiometers, can be used as automatic input units.

2.10. Boundary Conditions in Numerical and Mechanical Integration of Differential Equations

Both in mechanical integration of differential equations by the differential analyser, and in numerical integration whether by pencil-and-paper methods or by the use of one of the large automatic digital machines to be considered later, one meets in pronounced form a point which is scarcely mentioned in the conventional formal treatment of differential equations. This is the difference between the situation in which all the conditions which the solution has to satisfy are boundary conditions given at one point (usually one end) of the range of integration — these will be called one-point boundary conditions — and the situation in which some conditions are given at one point and others at another — these will be called two-point boundary conditions — or in which the solution y as a whole has to satisfy some integral relations, such as a normalising condition like $\int_0^1 y^2 dx = 1$ or relations of orthogonality to other functions.

The reason for the difference of emphasis on this point is that in the formal treatment a solution of a differential equation is thought of as a function, existing over the whole relevant range of the independent variable, which satisfies the equations at each point of the range. The purpose of the formal treatment is to find such a function; and the idea of the differential equation describing a *process* in which the independent variable increases and the dependent variable behaves in conformity with the equation is secondary. In mechanical or numerical solution, however, this aspect is most prominent, as it is just by following out such a process that a solution is evaluated.

Such a process has got to start from somewhere, with numerical values of all relevant quantities given there. If the conditions to be satisfied are one-point boundary conditions, the solution can start from the point where they are given, with everything known, and can proceed from there, normally without difficulties, except possibly those arising from singularities or of instability of the process of solution.

But when the conditions include two-point boundary conditions, or integral conditions on the solution as a whole, it is usually necessary to estimate one or more initial conditions at the point from which the solution is started, or perhaps one or more parameters in the equation itself, and to run a number of trial solutions with different values of these estimates until one satisfying the conditions is found or can be interpolated with sufficient accuracy.

Sometimes when the equation is linear a solution satisfying two-point boundary conditions can be obtained by evaluating a particular integral and a sufficient number of complementary functions, all satisfying the boundary conditions at one end of the range, and forming a linear combination satisfying the conditions at the other end of the range. But my experience has been that this is usually a purely formal possibility and not a practical one, because of the sensitiveness of the solution to the initial conditions, which would lead to the values of the final solution being evaluated as the small differences of two sets of large numbers. In such cases the use of trial solutions is preferable, and for non-linear equations it is the only way.

The use of trial solutions may mean that a large number of differential analyser runs are required for each useful solution obtained. On the other hand all these trial solutions refer to the same equation so that no changes of set-up are required in evaluating them; in most cases the only changes required between one trial solution and another will be changes of initial displacements of integrators.

2.11. Applications

The differential analyser evaluates solutions of differential equations without reference to the particular physical, chemical, technical, or other problems in which the equations occur. The range of problems to which it may make a contribution is correspondingly wide. It has been applied to calculations concerning the structures of atoms (45, 78) and of stars, to the motion of electrified particles in the magnetic field of the earth in interplanetary space (75) and to running times of railroad trains (56), to the effect of a time-lag on the performance of automatic control circuits (19, 20), to problems in non-linear electrical circuits (33, 60, 109), fluid motion (46), and chemical kinetics (see ref. 27, p. 83), to the oscillations of the atmosphere (111) and to a wide variety of other problems (see refs. 27, 109).

Chapter 3

THE DIFFERENTIAL ANALYSER AND PARTIAL
DIFFERENTIAL EQUATIONS

3.1. Introduction

There have been two main developments in connection with the differential analyser since the first instrument of this kind was constructed at M.I.T. One of these is concerned with the development of an improved, more flexible and more accurate, form of the instrument itself, as represented by the new differential analyser of Bush and Caldwell described shortly in §2.5. The other is concerned with an extension of the field of application of the instrument to partial differential equations; this has been carried out mainly by my group at the University of Manchester before and during the war.

The differential analyser will handle directly only *ordinary* differential equations. Before applying it to a *partial* differential equation, this must be replaced by a set of ordinary differential equations which are approximately equivalent to it. Thus the application of the differential analyser to a partial differential equation depends essentially on an approximation. But, as will be seen, it is possible to correct for the main part of the error involved in this approximation, and it is quite practicable to obtain results which, while still approximate, are fully accurate enough to be of practical value. The approximation involved here is similar to that involved in any application of a *digital* machine to the solution of differential equations, ordinary or partial.

It will be convenient to illustrate the procedure first for the simple partial differential equation of heat conduction in one dimension in a substance with uniform and constant thermal properties. In terms of reduced length- and time-variables x and t, this equation is

$$\frac{\partial \theta}{\partial t} = \frac{\partial^2 \theta}{\partial x^2} \tag{3.1}$$

where θ is the temperature.

This equation can be replaced by a set of ordinary differential equations by replacing the derivative with respect to either one of the independent variables by a finite difference. The resulting situation is considerably different according as it is the t-derivative or the x-derivative which is so replaced.

3.2. Replacement of the t-Derivative by a Finite Difference

Consider first the result of replacing the t-derivative by a finite difference (61). With an error term of order $(\delta t)^2$, we have

$$\left(\frac{\partial \theta}{\partial t}\right)_{t=t_0+\frac{1}{2}\delta t} = \frac{\theta(t_0 + \delta t) - \theta(t_0)}{\delta t}$$

and with an error term of the same order

$$\left(\frac{\partial^2 \theta}{\partial x^2}\right)_{t=t_0+\frac{1}{2}\delta t} = \frac{1}{2}\left[\frac{\partial^2 \theta(t_0 + \delta t)}{\partial x^2} + \frac{\partial^2 \theta(t_0)}{\partial x^2}\right].$$

Hence, neglecting terms of order $(\delta t)^2$, we can replace (3.1) by

$$\frac{\partial^2}{\partial x^2}\left[\theta(t_0 + \delta t) + \theta(t_0)\right] = \frac{2}{\delta t}\left[\theta(t_0 + \delta t) - \theta(t_0)\right]. \qquad (3.2)$$

If the temperature distribution at time t_0 is given, that is to say, $\theta(t_0)$ is given as a function of x, this is an *ordinary* differential equation for $\theta(t_0 + \delta t)$ as a function of x, that is to say, for the temperature distribution at time $t_1 = t_0 + \delta t$. Once this has been determined, it becomes the known initial temperature distribution for the interval t_1 to $t_2 = t_1 + \delta t$, and so on.

The procedure can be represented by fig. 22, in which the heavy lines represent the boundary of the field of integration in x and t, and the arrow indicates the direction in which the integration is carried.

Three points should be noticed at this stage:

(i) The only input necessary for the interval t_0 to t_1 is $\theta(t_0)$ as a function of x; $\partial^2 \theta(t_0)/\partial x^2$ is not required, for a two-fold integration of (3.2) gives $\theta(t_0+\delta t)+\theta(t_0)$, and from this and the input $\theta(t_0)$, $\theta(t_0+\delta t)$ can be derived.

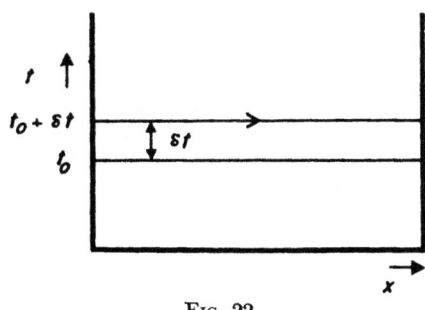

FIG. 22

(ii) The equations for the successive t-intervals are solved *successively*, and, since they differ only in the input function which is $\theta(t_0)$, $\theta(t_1)$, ... in succession, the same set-up can be used for them all.

(iii) The solution for each t-interval has to satisfy *one* boundary condition at *each* end of the x-range.

For a given point (x,t) in the field of integration, the value of θ calculated by this process depends not only on x and t, but on the time interval (δt) used in the integration. This can be expressed by writing this result as $\theta(t, x, \delta t)$. The required solution of the equation is $\theta(t, x, 0)$.

An analysis of the error term shows that in one interval the error in $\theta(t_0 + \delta t) - \theta(t_0)$ at any point x is of order $(\delta t)^3$. The number of intervals required to cover a given total time T is $T/(\delta t)$. So the aggregate error in a given total time T is of order $(\delta t)^2$. The leading term in this error can in principle be eliminated by covering the same total time T by two independent integrations with different interval lengths (δt), and extrapolating the results to an interval length $(\delta t) \rightarrow 0$, taking $\theta(t, x, \delta t)$ for given (t, x) to be linear in $(\delta t)^2$. This is an example of a process proposed some time ago by L. F. Richardson and called by him "h^2-extrapolation" (95). It is most convenient to take δt for one solution twice as large as for the other, say $\delta t = \Delta$ and $\delta t = 2\Delta$; and the most convenient form for the result is

$$\theta(t, x, 0) = \theta(t, x, \Delta) - \tfrac{1}{3}\left[\theta(t, x, 2\Delta) - \theta(t, x, \Delta)\right]. \tag{3.3}$$

The process is illustrated graphically in fig. 23. A fuller analysis of the error shows that in many cases this process eliminates not only the $(\delta t)^2$ term in the error but the $(\delta t)^3$ term also (see ref. 61).

In practice, it appears to be most convenient to take four intervals $\delta t = \Delta$ and two intervals $\delta t = 2\Delta$, to carry out the process of "h^2-extrapolation" on the results, and to continue the integration from the results so corrected.

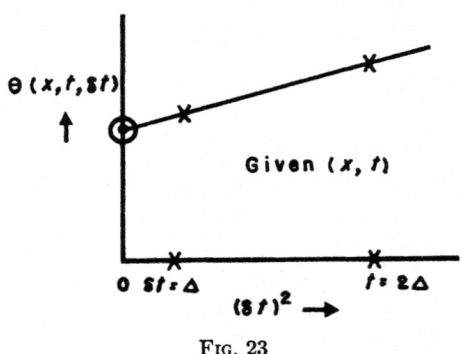

Fig. 23

3.3. Replacement of the x-Derivative by Finite Differences (32)

To replace the x-derivative by a finite difference, let the x-range be divided into a number of equal intervals δx, let $x_j = j\delta x$ be a typical point of division and let θ_j be the temperature there. Then the finite-difference approximation to $(\partial^2\theta/\partial x^2)$ at x_j is

$$\left(\frac{\partial^2\theta}{\partial x^2}\right)_{x=x_j} = \frac{\theta_{j+1} - 2\theta_j + \theta_{j-1}}{(\delta x)^2}$$

with an error term of order $(\delta x)^2$. By equation (3.1), this should be equated to the time derivative of θ_j, so that we get the following set of equations

$$\frac{\partial \theta_j}{\partial t} = \frac{\theta_{j+1} - 2\theta_j + \theta_{j-1}}{(\delta x)^2}. \tag{3.4}$$

The original proposal for a mechanism to handle the set of equations (3.4) was made to me by Dr. R. J. Sarjant and Dr. R. Jackson of the Research Department of Messrs. Hadfields Ltd. of Sheffield. They proposed to couple a series of integrators by a linkwork which should make the integrand ($\partial\theta/\partial t$) of any integrator depend on the outputs (θ) of that integrator and its neighbours in the way specified by equation (3.4). But the differential analyser could already do just this; there was no need to design and build a special instrument for the purpose.

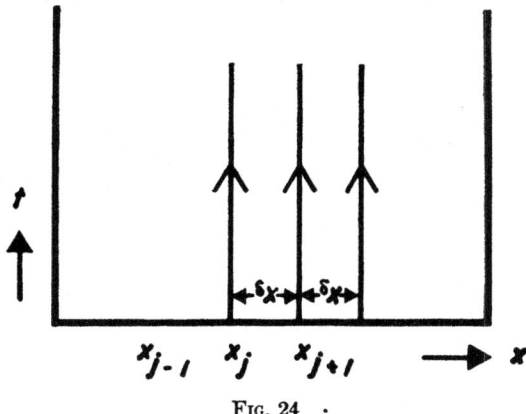

Fig. 24 ·

The integration process is illustrated diagrammatically in fig. 24 in which, as in fig. 22, the heavy lines indicate the boundary of the field of integration and the arrows indicate the direction in which the integration is carried.

If the surface conditions are such that the surface temperature is given as a function of t, then at a value of x_j one interval in from a surface, θ_{j+1} or θ_{j-1} in equation (3.4) is a surface temperature, and is input as a function of t. Surface conditions involving the surface heat flux can also be handled, though not quite so easily (see ref. 32).

The following four points should be noted:

(a) The set of equations (3.4) has to be solved *simultaneously*; this makes considerable demands on differential analyser capacity, unless an adequate approximation is given by use of only a quite small number of x-intervals.

(b) There is only one boundary condition for each equation; this is given by the initial temperature at each point x_i, and gives no difficulty; the boundary conditions in x are input continuously as the solution proceeds, and no trial-and-error process is required in order to satisfy them.

(c) There is no difficulty in principle to extending the method to heat conduction in two or three dimensions in space; there is, however, a possible practical difficulty in the amount of equipment required, since one integrator is required for each point of the space-field.

(d) The calculated solution is a function of δx as well as of x and t, and the leading term in the error at any (x,t) is of order $(\delta x)^2$. This can be eliminated by an application of the method of "h^2-extrapolation" already explained.

Radial heat flow in a cylinder or sphere is formally similar to heat flow in one dimension, in that the temperature depends on the time and one space-coordinate. For a *solid* cylinder, let the radius be divided into an integral number of intervals δr and let $r_j = j\delta r$. Then the finite-difference form of

$$\frac{\partial^2\theta}{\partial r^2} + \frac{1}{r}\frac{\partial\theta}{\partial r} = \frac{\partial\theta}{\partial t} \tag{3.5}$$

is

$$\frac{\partial\theta_j}{\partial t} = \frac{1}{(\delta r)^2}\left[\left(1 + \frac{1}{2j}\right)\theta_{j+1} - 2\theta_j + \left(1 - \frac{1}{2j}\right)\theta_{j-1}\right] \tag{3.6}$$

for $j \neq 0$, and

$$\frac{\partial\theta_0}{\partial t} = \frac{4}{(\delta r)^2}(\theta_1 - \theta_0). \tag{3.7}$$

For a *hollow* cylinder it may be convenient to use an appropriate group of the equations (3.6), or it may be better to transform from r to $\log r$ as space-variable; then the equations reduce to the one-dimensional form but with the effective diffusivity a continuous function of $(\log r)$.

3.4. Discussion of the Two Methods

As a simple example, the equation of heat conduction in one dimension in a uniform substance with constant thermal properties, and in which there is no generation of heat, has so far been considered. But neither method is restricted to conduction in a substance with constant thermal properties. In one dimension, the equation of conduction in a substance of specific heat c, density ρ, and conductivity K, in which heat is being generated at a rate G, is

$$c\rho\frac{\partial\theta}{\partial t} = \frac{\partial}{\partial x}\left(K\frac{\partial\theta}{\partial x}\right) + G.$$

It is convenient to take the equation in a form in which first derivatives in x do not occur, and this can be done by taking

$$\psi = \int_{\theta_0} (K/K_0)d\theta$$

as dependent variable; this can be regarded as the temperature measured on a distorted scale. Then, with $K/c\rho = D$, the equation becomes

$$\frac{\partial \psi}{\partial t} = D(\psi) \left[\frac{\partial^2 \psi}{\partial x^2} + \frac{G}{K_0} \right] \tag{3.8}$$

The rate of generation of heat G may be given as a function of x, t or θ; in any case the methods of §§ 3.2, 3.3 are both applicable.

The main practical difference between the two methods is that in the first (finite intervals in t), each function $\theta(t)$ calculated has to satisfy two-point boundary conditions, whereas in the second (finite intervals in x) the conditions are all given at a single value of the independent variable. The great difference between these two situations has already been emphasized in §2.10. In the first method, a large number of trial solutions will usually be required for each interval (δt), before one is found satisfying the boundary conditions; an example is given in fig. 25. A similar set of trial solutions has to be run for each (δt). Moreover, the smaller the value of (δt) taken, in the interests of accuracy of the solution, the more sensitive the trial solutions are to the starting conditions. Thus the evaluation

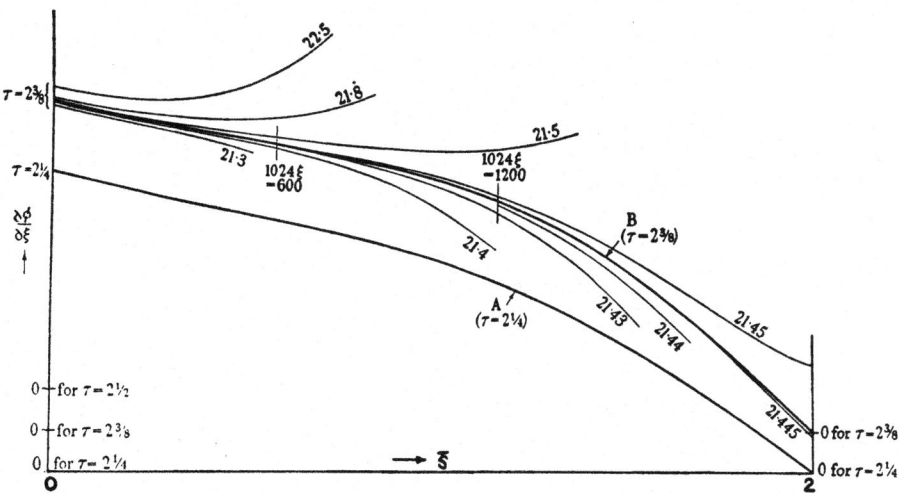

Fig. 25. A step in the solution of the equation $\dfrac{\partial \phi}{\partial \tau} = \dfrac{\partial^2 \phi}{\partial \xi^2} + \beta e^{\phi}$ (see ref. 25); ϕ, τ, ξ are reduced measures of θ, t, x. Conditions $\phi = 0$ at $\xi = 0$, $\partial \phi / \partial \xi = 0$ at $\xi = 2$. The thin curves show trial solutions for the interval $\tau = 2\frac{1}{4}$ to $2\frac{3}{8}$ and each is labelled with the value of $\partial \phi / \partial \xi$ at $\xi = 0$.

of a whole solution as a function of x and t may easily require some hundreds of separate runs of the differential analyser, and is not a project to be undertaken lightly. On the other hand with the second method, a single run, or two runs with different values of (δx), are all that is required, so that the time required to evaluate a whole solution is some hundreds of times less than with the first method.

However, in the first method the equations for the successive time intervals can be solved successively, so that the demands on equipment are limited to those required to evaluate solutions of the single equation (3.2), or of the corresponding finite-difference form of equation (3.8). These will be fairly small. Exactly what they are will depend on the nature of the functions $D(\psi)$ and G, and whether these can be generated or have to be fed in from input tables; they are not likely to be more than two or three integrators and two or three input tables.

But in the second method the equations for the different quantities θ_j have to be integrated *simultaneously*. For the simple equations (3.4), derived from the simplest form (3.1) of the equation of non-steady heat conduction, this requires one integrator for each point x_j of the space-range. And for the corresponding finite-difference form of the more general equation (3.8) with diffusivity a function of temperature, even with no generation of heat, it requires two integrators for the integration, and, usually, an input table to give $D(\psi)$ as a function of ψ, for *each* point x_j. A number of input tables are wanted, since although D is the same function of ψ at all points, it is required to feed in *simultaneously* the values of D at all the various values of ψ_j, and this cannot be done continuously from a single input table. Similarly, if there were a generation of heat G in the body of the material, an input table would usually be needed to give this at each value of x_j. It is of course possible that the variation of D with ψ, or of G with those variables on which it depends, could be generated by one or more integrators; but the integrator set-up for this purpose would have to be repeated for each point x_j.

Thus this method makes considerable demands on differential analyser capacity, particularly in integrators and, perhaps, input tables; the set-up details show that it also makes a considerable demand on adding units. On the other hand when it can be used, its advantages over the first method are so very marked that it should always be used when adequate accuracy can be achieved with the comparatively small number of points x_j which must be taken to keep the equations within the available differential analyser capacity.

However, trial solutions have shown that in many cases of practical interest, results of useful accuracy can be obtained with quite a small number of intervals δx or δr. For instance four or three, or even two, intervals in the half-thickness of a symmetrically heated plate, or in the radius of a symmetrically heated cylinder, are often adequate. This means that it is practicable, with available differential analysers, to deal with simple cases in which the temperature depends on two space variables as well as on the time. This is a further advantage, in principle, of this method; it does not seem practicable to make a corresponding extension of the first method.

A feature of the situation which makes either method practicable is that *the field of integration is open in the time direction;* there are initial conditions at the starting time $t = 0$, but no conditions to be satisfied when the solution reaches some later time $t = T$. This is an essential feature. If some conditions had to be satisfied at $t = T$, it would be necessary to estimate a set of quantities, or a whole function, at $t = 0$, and to run trial solutions with different estimates until one was found satisfying the conditions at $t = T$; the amount of work involved would make the process impracticable.

The difference between a closed field of integration and one which is open in one direction is similar to the difference, in ordinary differential equations, between two-point and one-point boundary conditions which has already been emphasized (§ 2.10). The form of the boundary conditions is often related to the character of the equation itself, namely to which of the types named "elliptic," "parabolic," or "hyperbolic" it belongs. An equation of "elliptic" type, for example Poisson's equation

$$\frac{\partial^2 \theta}{\partial x^2} + \frac{\partial^2 \theta}{\partial y^2} = f(x,y)$$

is usually associated with closed-boundary conditions. For such equations use of the differential analyser is inappropriate. A more suitable technique is the use of some form of Southwell's "relaxation" method (101, 102).

"Parabolic" equations, for example the equation of heat conduction

$$\frac{\partial \theta}{\partial t} = D(\theta) \frac{\partial^2 \theta}{\partial x^2}$$

and "hyperbolic" equations such as the wave equation

$$a^2 \frac{\partial^2 \theta}{\partial x^2} - \frac{\partial^2 \theta}{\partial t^2} = 0$$

are usually associated with boundary conditions on a boundary open in one co-ordinate, which physically is often the time co-ordinate. Considerable success has been achieved in applying the differential analyser to equations of "parabolic" type; some examples will be considered in the next section. A few attempts have been made to apply it to equations of "hyperbolic" type, with only partial success, but as this is a field of application of the differential analyser which seems to deserve further exploration, and on instruments of capacity greater than those so far available in England, a survey of this work is given here in §§3.6-3.9.

3.5. Examples of Application to Equations of "Parabolic" Type

Most of the work that has been done has been on the equation of heat conduction, and most of this has been concerned with heat conduction in steel, taking account of the variation of the thermal properties with temperature (32, 71).

Figure 26 shows a comparison of calculated and observed temperature in a cylindrical bar, the experiment being carried out under conditions designed to give a good approximation to axially-symmetrical heating uniform along the bar, so that the heat flow was mainly radial. The surface and centre temperatures were measured by thermocouples, the surface temperature as a function of the time, and the thermal properties as functions of temperature, being the only data used in the calculation of the central temperature.

In the calculation, the abnormal thermal properties of steel in the neighbourhood of 715° C., associated with the $\alpha - \gamma$ transformation of the iron lattice, were represented by a latent heat of transformation at this temperature; actually the transformation is a more complex phenomenon and at high rates of change of temperature such as those occurring in this experiment, it occurs over a range of temperature. This accounts for the departure of the calculated central temperature from that observed, in this temperature region.

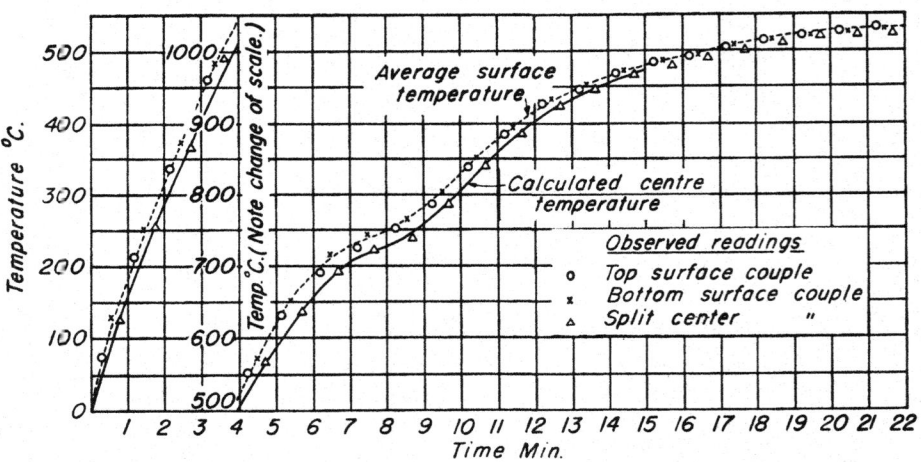

Fig. 26

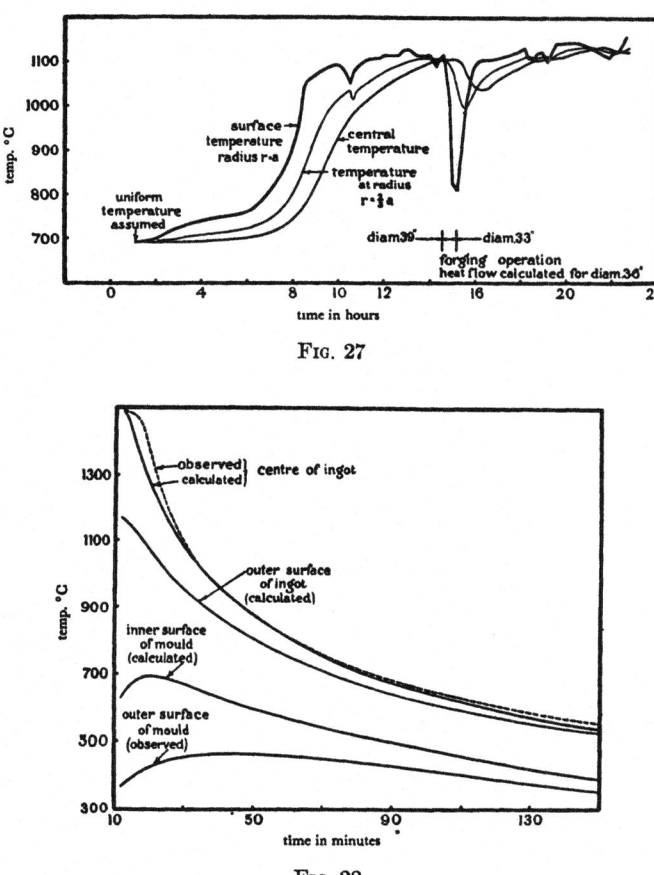

Fig. 27

Fig. 28

Figure 27 shows the results of calculation of the internal temperature of a large forging during forging and reheating operations.

Figure 28 shows a result of the most ambitious application of the differential analyser to problems of this kind, one concerning the cooling of an ingot in a mould. When molten steel is poured into a mould, the outer surface of the cast ingot soon solidifies, and then as the mould heats up and expands and the ingot cools down and contracts, a gap is formed between the outside of the ingot and the inside of the mould. Heat transfer is then by conduction through the material of the ingot and through the mould wall, and by radiation across the gap between them. It was possible to set up on the differential analyser the equations of radiative heat transfer according to Stefan's law, as well as the equations of heat conduction in finite-difference form, and so obtain solutions of this rather complex heat-transfer problem.

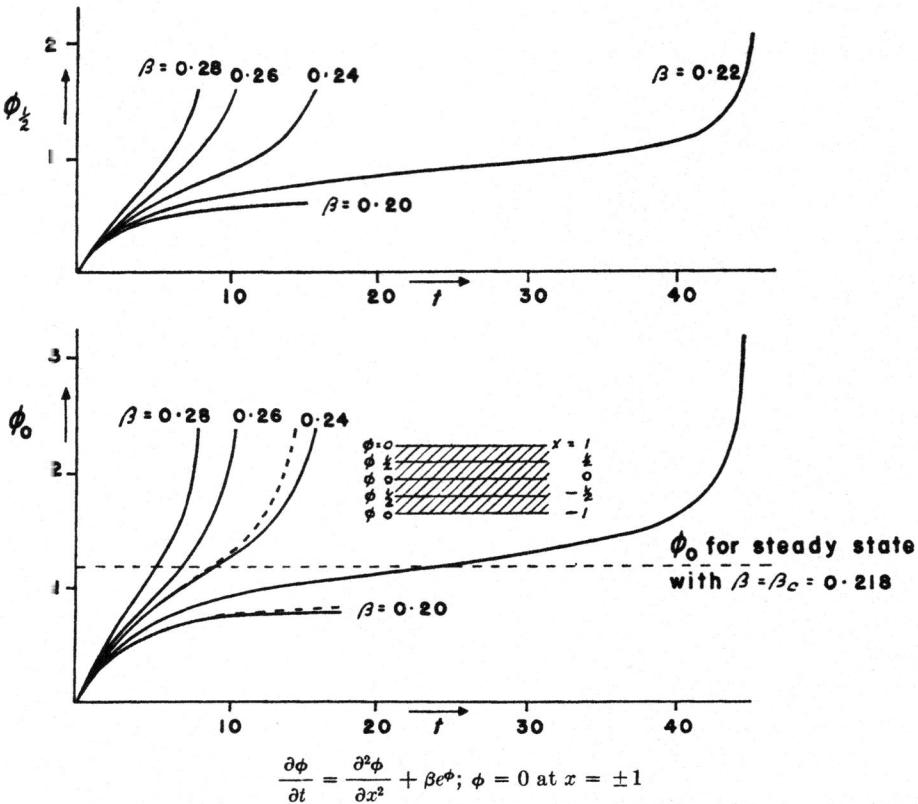

$$\frac{\partial \phi}{\partial t} = \frac{\partial^2 \phi}{\partial x^2} + \beta e^\phi; \quad \phi = 0 \text{ at } x = \pm 1$$

Fig. 29. Calculated time variation of central temperature in a slab in which rate of generation of heat increases exponentially with the temperature; surfaces held at zero temperature. ϕ = reduced temperature; rate of generation of heat = βe^ϕ.

Figure 29 shows some results calculated for the temperature distribution in a substance in which there is a generation of heat increasing exponentially with the temperature, so that the equation of one-dimensional heat conduction, in reduced variables, is

$$\frac{\partial \phi}{\partial t} = \frac{\partial^2 \phi}{\partial x^2} + \beta e^\phi.$$

For given boundary conditions, there is a critical value of β above which this equation has no steady state solution; for the conditions $\phi = 0$ at $x = \pm 1$, to which the solutions shown in fig. 29 refer, this critical value is $\beta = 0.218$. This equation is of great interest in connection with the thermal behaviour of solid dielectrics in alternating electric fields. For many such substances the power loss, for a given amplitude of the electric stress, increases approximately exponentially with the temperature (41). The calculated solutions show good general agreement with the main features of the actual thermal behaviour of such substances, with the final catastrophic onset of thermal instability, and the long period which elapses before the temperature rise becomes catastrophic if the stress is only slightly above the critical value. Some solutions of this equation were first obtained by application of the method of §3.2 (25); later, solutions were obtained using the method of §3.3 (49), and showed very markedly the advantages of the latter method when adequate differential capacity is available.

The equation of the laminar boundary layer in fluid dynamics, though third order, is formally of hyperbolic character, and the boundary conditions are of the corresponding "open" type, in that (to the approximations of boundary layer theory) the solution has not got to satisfy any conditions at a distance downstream. An application of the method of §3.2 to this equation was successful though very lengthy (48).

3.6. Hyperbolic Equations

The only work on the application of the differential analyser to hyperbolic equations of which I know is that carried out in England during the war. On account of the pressure of war conditions, very little preliminary exploratory work on *simple* hyperbolic equations was done, and the attempts to treat more complicated equations were only partly successful. This work is mentioned here, since as far as it goes it suggests that this field of application of the differential analyser deserves further study, and with instruments of greater capacity than those at present available in England.

The simplest hyperbolic equation is the wave equation in a uniform medium in which the velocity of sound is constant. In terms of reduced variables x and t, this equation can be written

$$\frac{\partial^2 \theta}{\partial t^2} = \frac{\partial^2 \theta}{\partial x^2} \tag{3.9}$$

Normally the boundary conditions are given on the same sort of open boundary as for the equation of heat conduction, and it would appear practicable to replace either derivative by a finite difference. The general character of the resulting situation is much the same as in the case of the equation of heat conduction. If the t-derivative is replaced by a finite difference, we get a set of equations which can be solved *successively* for the successive time intervals, but for each t-interval we have to satisfy two-point space-boundary conditions, and so have to evaluate a large number of trial solutions for each useful result obtained. Whereas if the

x-derivative is replaced by a finite difference, we have one-point boundary conditions in t, and the boundary conditions in x are satisfied continuously as the solution progresses, without any process of trial-and-error; but the set of equations for the various points x_j have to be solved *simultaneously*, and this leads to demands on differential analyser capacity which are still more severe than for the equation of heat conduction, since the equation for each θ_j is now second-order in t.

Neither of these methods has been tried on the simple equation (3.9), but both were tried on a more elaborate equation, or rather system of equations, namely those for the propagation of a spherical compression-wave of large amplitude, under anisentropic conditions, and neither led to a practicable procedure. It should be said, however, that attempts to treat this problem by other methods led to the conclusion that it was not easily amenable to any method of treatment, and that the evaluation of a solution is a major computing operation.

3.7. Use of "Characteristics"

A different approach to the treatment of hyperbolic equations was successful in one application, and it seemed that only the restrictions set by the capacity of the differential analysers available in England at the time prevented its success in application to a more intricate situation.

In the theory of hyperbolic equations, two sets of curves in the plane of integration, called "characteristics," play an important part. If the equation is

$$H\,\frac{\partial^2\theta}{\partial x^2} + 2K\,\frac{\partial^2\theta}{\partial x\partial y} + L\,\frac{\partial^2\theta}{\partial y^2} = M \qquad (3.10)$$

where H, K, L, and M may be functions of any of the variables $\partial\theta/\partial x$, $\partial\theta/\partial y$, θ, x and y, the characteristics are curves in the (x,y) plane whose slope at any point is given by

$$H\left(\frac{dy}{dx}\right)^2 - 2K\left(\frac{dy}{dx}\right) + L = 0. \qquad (3.11)$$

For a hyperbolic equation, the roots of this equation are real and distinct, so that there are two sets of characteristics, and one characteristic of each set passes through each point (x,y). On a characteristic, equation (3.10) becomes

$$H\,\frac{dy}{dx}\,d\left(\frac{\partial\theta}{\partial x}\right) + L\,d\left(\frac{\partial\theta}{\partial y}\right) = M\,dy. \qquad (3.12)$$

If H, K, L are independent of θ and its derivatives, then the characteristics *depend only on the equation and not on the particular solution con-*

sidered; the evaluation of the characteristics and of the solution can then be carried out *successively,* the characteristics being determined first by solution of (3.11) and the solution of (3.12) by integration along the characteristics being carried out subsequently. But if one or more of H, K, L depend on θ or its first derivatives, the characteristics depend on the particular solution evaluated, and the determination of the characteristics and of the solution have to be carried out *simultaneously.*

The equation for the axially symmetrical, isentropic, motion of a compressible inviscid fluid, around a solid of revolution, is

$$\left(1 - \frac{u^2}{a^2}\right)\frac{\partial^2\phi}{\partial x^2} + \frac{2uv}{a^2}\frac{\partial^2\phi}{\partial x \partial y} + \left(1 - \frac{v^2}{a^2}\right)\frac{\partial^2\phi}{\partial y^2} + \frac{v}{y} = 0 \quad (3.13)$$

where x and y are the axial and radial co-ordinates, ϕ the velocity potential, $u = \partial\phi/\partial x$, $v = \partial\phi/\partial y$, and a is the local velocity of sound, given by

$$a^2 = a_0{}^2 - \tfrac{1}{2}(\gamma - 1)(u^2 + v^2).$$

This equation is hyperbolic where the flow is supersonic $(u^2 + v^2 > a^2)$. What is wanted is the velocity (u, v), rather than ϕ itself, as a function of (x,y).

It is convenient to write $dy/dx = 1/\mu_1$, $1/\mu_2$ for the two roots of (3.10), from which it follows that $H = L\mu_1\mu_2$. Then for one set of characteristics, which will be called set 1, namely that for which

$$dx = \mu_1\, dy, \qquad\qquad\qquad (3.14a)$$

it follows from (3.12) that

$$\mu_2\, du + dv = (\dot{M}/L)dy \qquad\qquad\qquad (3.15a)$$

and for the other set (set 2)

$$dx = \mu_2\, dy \qquad\qquad\qquad (3.14b)$$

and

$$\mu_1\, du + dv = (M/L)dy. \qquad\qquad\qquad (3.15b)$$

Equations (3.14) define the shapes of the characteristics, and equations (3.15) define the velocity distribution along them. It is convenient to use the term "corresponding points" for points on the various characteristics of one set where they are cut by a single characteristic of the other set.

Now suppose we try to solve these equations by integrating successively along a sequence of characteristics of one set, say of set 2, using finite differences between successive characteristics of this set. Let A and B be two characteristics of this set (see fig. 30). Then the finite difference forms of (3.14a) and (3.15a) are

$$x_B - x_A = \tfrac{1}{2}(\mu_{1A} + \mu_{1B})(y_B - y_A) \tag{3.16}$$

and

$$\tfrac{1}{2}(\mu_{2A} + \mu_{2B})(u_B - u_A) + (v_B - v_A) = \tfrac{1}{2}[(M/L)_A + (M/L)_B](y_B - y_A). \tag{3.17}$$

For any point x_B, y_B on characteristic B, the quantities in equation (3.17) referring to characteristic A are to be taken at the corresponding point (x_A, y_A) given by (3.16). Given the characteristic A and the velocity distribution along it, that is, given x_A, u_A, and v_A as functions of y_A, equations (3.11) (which gives μ_1, μ_2), (3.14b), (3.15b), (3.16), (3.17) give the shape of characteristic B and velocity distribution along it.

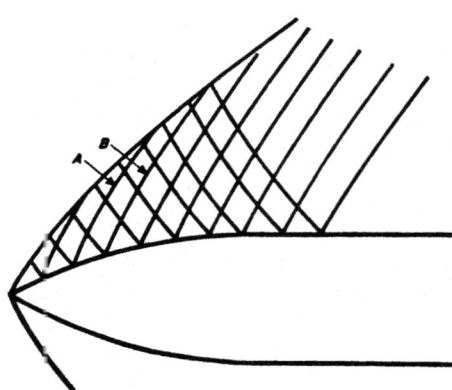

For a slender body, a useful approximation for some purposes is given by taking u as constant and v as zero in the coefficients of the second derivative terms in (3.13). In this approximation, the characteristics are independent of the particular solution, and are indeed straight lines, and the application of the differential analyser to the solution of equations (3.15b) and (3.17) was quite successful. This application was carried out by the differential analyser group at the Mathematical Laboratory of the University of Cambridge (ref. 27, p. 106).

FIG. 30. Characteristics for supersonic flow past a shell.

Previously, an attempt had been made, by my group at Manchester, to deal with the more general, but considerably more involved, situation in which the characteristics depend on the solution being evaluated. With the capacity available, it was not possible to set up and solve simultaneously the whole set of equations (3.11), (3.14b), (3.15b), (3.16), (3.17) giving the shape of characteristic B and the velocity distribution along it. An iterative process was tried, in which the shape of a characteristic was first estimated and the velocity distribution along such a characteristic evaluated on the differential analyser by solution of (3.15b) and (3.17); this was used to evaluate H, K, L and hence better values of μ_1, u_2 from (3.11) and a revised shape for the characteristic obtained, and so on. The process turned out to be very long and tedious; further, the part of the work for which the differential analyser was used was quite a small fraction of the whole. The attempt to use the differential analyser for this work was therefore abandoned for other work on which it could be used more efficiently.

From this experience, it cannot be said definitely whether it would have been practicable to carry out solutions by this method if there had been available a differential analyser of adequate capacity to handle simultaneously the whole set of equations (3.11), (3.14b), (3.15b), (3.16), and (3.17), or whether the method has other weaknesses which would vitiate it as a practicable process. The impression of those who actually took part in the work is that the former is probably the case, and it would be very interesting to carry out further work on this problem on a differential analyser of adequate capacity.

3.8. Another Application of Characteristics

Another possible method is to use finite differences with respect to one of the independent variables, but to make use of the characteristics to express the relations between quantities at the beginning and end of this interval. This treatment leads to a remarkable differential analyser set-up

For sound-waves of finite amplitude under conditions in which the density is the same function of the pressure at all points of the fluid and all times, there is a standard transformation, due to Riemann (74, 96), which, for waves in one dimension, gives the following:

If u, a are the gas velocity and the velocity of sound at any point, let

$$f(\rho) = \int a \, d\rho/\rho$$
$$P = f(\rho) + u$$
$$Q = f(\rho) - u$$

then

$$\frac{dP}{dt} = 0 \text{ for a point travelling with velocity } \frac{dx}{dt} = u + a,$$

and

$$\frac{dQ}{dt} = 0 \text{ for a point travelling with velocity } \frac{dx}{dt} = u - a.$$

The curves $dx/dt = u \pm a$ are the characteristics, in the (x,t) plane, of the equation of propagation of waves of finite amplitude.

For spherical waves of finite amplitude, the characteristics are the same (x being replaced by the radial co-ordinate r), but the equations $dP/dt = 0$, $dQ/dt = 0$ are replaced respectively by

$$\frac{dP}{dt} = -\frac{2au}{r}, \qquad \frac{dQ}{dt} = -\frac{2au}{r}.$$

Now suppose P and Q (and hence $f(\rho)$ and u) are known as functions of r at time t_0 (see fig. 31) and consider a later time $t_1 = t_0 + \delta t$. Let A be a point in the (r,t) plane at $t = t_1$, and let the characteristics through A cut the line $t = t_0$ in B and C. Then we have approximately

$$
\left.
\begin{aligned}
r_A - r_B &= \tfrac{1}{2}\left[(u+a)_A + (u+a)_B\right]\delta t \\
r_A - r_C &= \tfrac{1}{2}\left[(u-a)_A + (u-a)_C\right]\delta t \\
P_A - P_B &= -\left[(au/r)_A + (au/r)_B\right]\delta t \\
Q_A - Q_C &= -\left[(au/r)_A + (au/r)_C\right]\delta t \\
2u_A &= P_A + Q_A \\
2f(\rho_A) &= P_A - Q_A
\end{aligned}
\right\} \quad (3.18)
$$

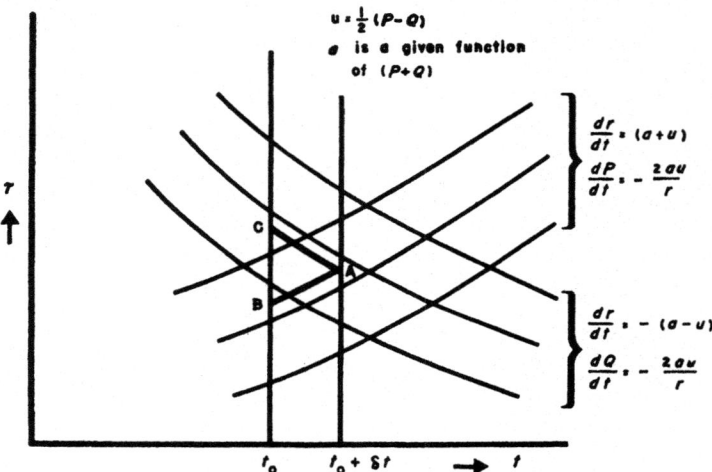

FIG. 31. Characteristics for spherical sound waves of finite amplitude.

and a is a definite function of $f(\rho)$. Given u and a as functions of r at time t_0, these equations are adequate to determine u and a as functions of r at time t_1. The third and fourth of equations (3.18) are most conveniently used in the forms

$$
[P + (au/r)\delta t]_A = [P - (au/r)\delta t]_B
$$
$$
[Q + (au/r)\delta t]_A = [Q - (au/r)\delta t]_C.
$$

A differential analyser set-up is given in fig. 32; it is remarkable for involving no integrators, only a rather intricate interconnection of input tables and the multiplier-divider unit, for producing a continuous solution of this set of non-linear algebraical equations as r_A varies. It has never been used, but provides another application of the differential analyser which would probably be worth further exploration.

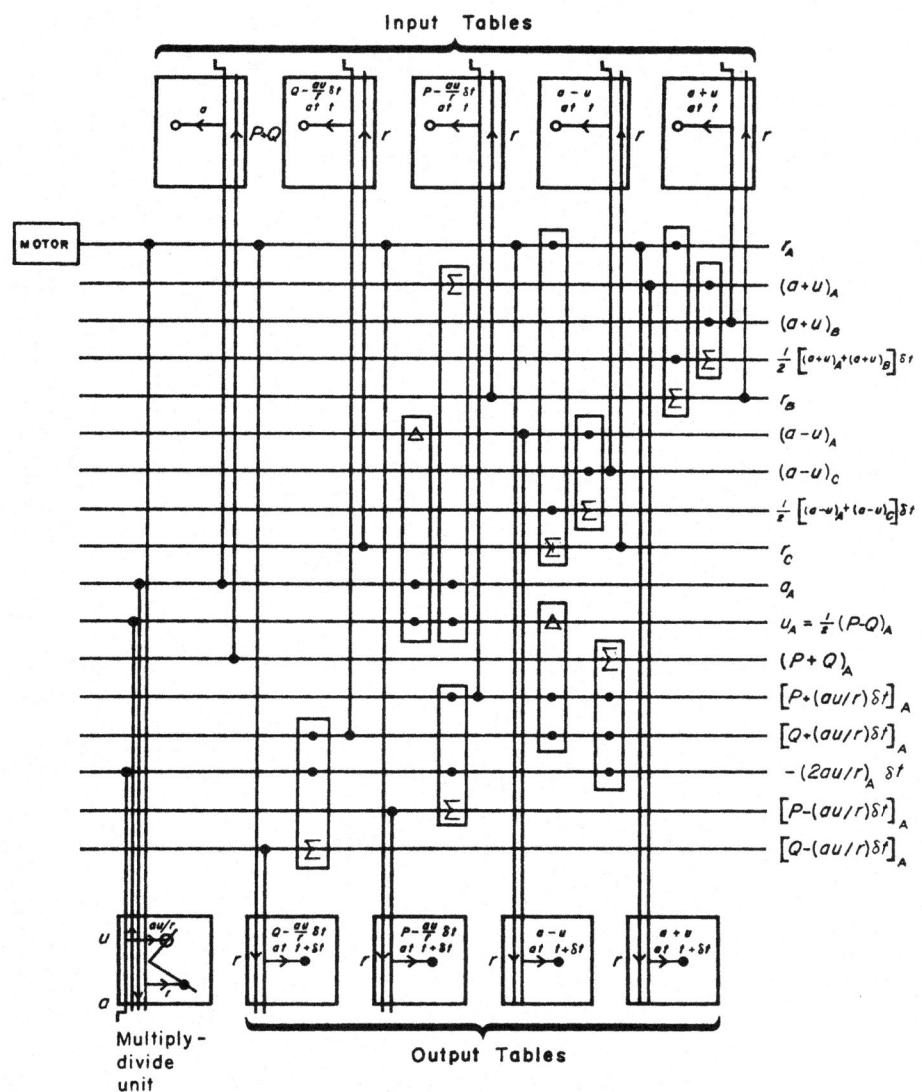

Fig. 32. Spherical sound waves of finite amplitude.

Chapter 4

SOME OTHER INSTRUMENTS

4.1. Introduction

A considerable number of instruments, in addition to the differential analyser, have been developed for various purposes; the present chapter is little more than a summary, indicating the main principles of some of them and giving references to fuller accounts of their construction and use. (For other references see 67, 86.)

As already mentioned, a single instrument is usually restricted to a single kind of calculation, so that one instrument may be used for solution of sets of simultaneous linear algebraic equations, another for harmonic analysis, and another for the evaluation of integrals as a function of a parameter. For carrying out any one kind of calculation, various instruments, mechanical or electrical, may be available, and instruments may be classified according to the kind of calculation for which they are suited, or according to the kind of mechanical or electrical means by which it is carried out. The former emphasizes the function, the latter the structure, of the instruments, and, in accordance with the balance to be adopted in discussing machines (see §5.2), the instruments will here be considered primarily from the point of view of function, that is to say, according to the kind of calculation they carry out.

4.2. Solution of Simultaneous Linear Algebraic Equations

Various instruments have been developed for the solution of simultaneous algebraic equations. The principle of an a.c. electrical instrument, due to Mallock (77), is shown in fig. 33. In this, the unknowns $x_j, \ldots$ are represented by the magnetic fluxes in various transformer cores, and the coefficients a_{ij} by the numbers of turns on the various cores. On each core there is a winding S which can be connected to an a.c. source, and another which can be connected to an a.c. voltmeter. A representative equation

$$\Sigma_j a_{ij} x_j = b_i \tag{4.1}$$

is set up as

$$\Sigma_j a_{ij} x_j - b_i u = 0$$

by taking a_{ij} turns on the core in which the flux is to measure x_j, and connecting up these turns, for a given i, in series in a closed circuit with b_i turns on another core, which will be called U. When all the equations have been set up in this way, an external voltage is applied to the S winding on one of the cores and adjusted so that the flux u in the core U

has some convenient value, which is taken as the unit of measurement; then the fluxes in the other transformers, measured in terms of this unit, give the values of the unknowns. This instrument can also be used to evaluate linear forms and determinants, and to obtain characteristic values λ and corresponding solutions of equations of the form

$$\Sigma_j \left(a_{ij} - \lambda \delta_{ij} \right) x_j = 0. \tag{4.2}$$

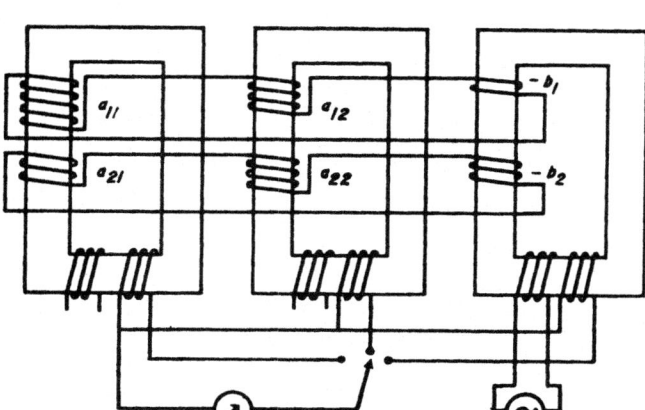

$$a_{11}x_1 + a_{12}x_2 - b_1 u = 0$$
$$a_{21}x_1 + a_{22}x_2 - b_2 u = 0$$

FIG. 33. Principle of Mallock's a.c. instrument for solution of simultaneous linear algebraic equations.

In the instrument as actually constructed there is capacity for handling 10 equations for 10 unknowns. The number of turns representing any coefficient can be set to any integral value from 0 to 1000, with sign.

On account of the limitations of any instrument, the set of values $x_j = \xi_j$ obtained as a solution will be only approximate; and, if the coefficients in the equation of which the solution is required are given to more than three figures, there is a further approximation involved in replacing these coefficients by three-figure numbers in the instrument. But unless the equations are so ill-conditioned that this approximation in the coefficients affects the solution considerably, a solution to any required accuracy can in principle be obtained in the following way. From an approximate solution ξ_i, the quantities

$$c_i = b_i - \Sigma_j a_{ij} \xi_j,$$

with the correct values of a_{ij}, are evaluated by other, more exact, means;

for example with a desk machine. The corrections $(x_j - \xi_j)$ to the approximate solution ξ_j satisfy the equations

$$\Sigma_j a_{ij} (x_j - \xi_j) = c_i \qquad (4.1a)$$

with the same coefficients as those of the original equations (4.1), and approximate values of these corrections can be found in the same way. This procedure can be repeated as often as required. It is not, of course, restricted to this instrument, but is a standard procedure in various processes of evaluating solutions of simultaneous equations.

In another form of instrument (1, 42) the x_j's are represented by the voltage outputs of a set of amplifiers, and the coefficients a_{ij} by the settings of resistive potential dividers. The output from a network producing a voltage representing $(\Sigma_j a_{ij} x_j - b_i)$ is fed back to the amplifier of which the output represents x_i. In a stable state, the outputs of the amplifiers satisfy the equations (4.1). (If the gain of the amplifiers is not large, a modification has to be made in the settings of the potential dividers representing the diagonal coefficients a_{ii}). Such an instrument can also be used to find the characteristic values λ and corresponding solutions of the set of equations (4.2). The stability of such a system of coupled feedback amplifiers has been investigated by Goldberg and Brown (42). Another study of feedback methods in the instrumental solution of simultaneous equations has been given by Taylor and Thomas (103).

Another form of a.c. instrument for linear simultaneous equations is the network analyser (66, 73, 89). This instrument was originally intended as an analogue in the simplest form of a model, in this case a model of an electrical power transmission system; it has a number of adjustable connections and circuit elements, so that it can be set up to represent different transmission systems, but it is not restricted to networks representing such systems. A number of these instruments have been built in the United States. One has recently been completed in England, and at least one other is under construction.

4.3. The Isograph

The isograph is the name given to an instrument for locating complex roots of polynominal equations, which was built by Bell Telephone Laboratories (29, 83).

If $\qquad f(z) = a_0 z^n + a_1 z^{n-1} + \ldots + a_n, \qquad a_k = |a_k| e^{i\gamma_k}$
then on a circle $z = re^{i\theta}$

$$\left. \begin{aligned} \operatorname{Re} f(z) &= A_0 \cos (n\theta + \gamma_0) + A_1 \cos [(n-1)\theta + \gamma_1] + \ldots \\ \operatorname{Im} f(z) &= A_0 \sin (n\theta + \gamma_0) + A_1 \sin [(n-1)\theta + \gamma_1] + \ldots \end{aligned} \right\} \quad (4.3)$$

where $A_k = |a_k| r^{n-k}$.

The instrument evaluates the quantities (4.3) mechanically and draws the locus described by the point representing $f(z)$ in its complex plane (see fig. 34). The quantities $k\theta$ $(1 \leqslant k \leqslant n)$ are obtained as rotations by mechanical gearing, and the quantities $A_k \cos (k\theta + \gamma_{n-k})$, $A_k \sin (k\theta + \gamma_{n-k})$ as linear displacements by resolving mechanisms. The addition of the terms in the expressions (4.3) is also carried out mechanically.

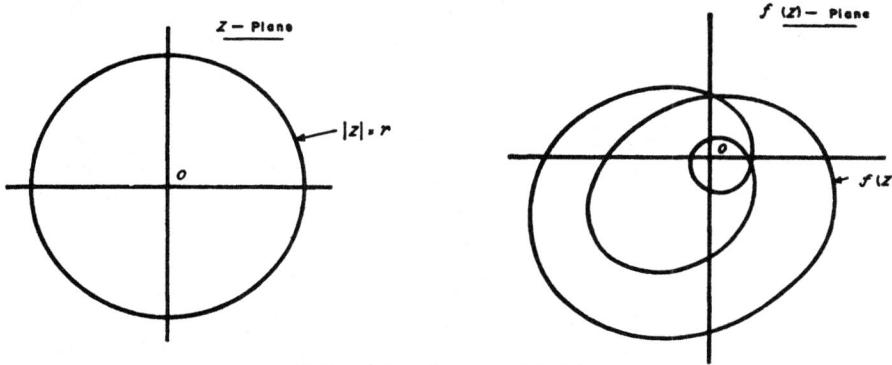

$f(z) = 0$ has 3 roots with $|z| < r$

Fig. 34

The value of r is determined by trial and error. An indication of the way in which to alter a trial value of r is provided by the theorem, in complex variable theory, that if z is taken around a closed curve, enclosing n roots of $f(z) = 0$, in the z plane, the curve described by $f(z)$ in its complex plane encircles the origin n times (see fig. 34). Thus if for a given value of r, the output curve drawn by the isograph encircles the origin n times, it follows that $f(z) = 0$ has n roots with $|z| < r$.

An electrical form of this instrument, using electrical resolvers, multipliers and adders, was planned in England towards the end of the war, but was not constructed, largely on account of inadequate supply of components.

4.4. Fourier Synthesisers

In the process of working out the results of X-ray analysis of crystal structures, it is necessary to evaluate sums of the form

$$\Sigma_{hk} F_{hk} \cos (hx + ky), \quad \Sigma_{hk} F_{hk} \sin (hx + ky) \qquad (4.4)$$

$$\Sigma_{hkl} F_{hkl} \cos (hx + ky + lz), \quad \Sigma_{hkl} F_{hkl} \sin (hx + ky + lz) \qquad (4.5)$$

for a number of sets of values of (x,y) or (x,y,z), the sums being over integral values of (h,k,l). The evaluation of the double and triple sums can be reduced to the evaluation of a number of single sums of the form $\Sigma_h A_h \cos h\xi$, $\Sigma_h A_h \sin h\xi$. Such sums are evaluated as continuous functions of ξ by the isograph in the course of its operation, but in the structure-analysis context it is usual to aim at calculating the sums for discrete values of (x,y,z), usually at 3° or 6° intervals.

An a.c. electrical instrument intended specifically for use in structure-analysis work has been designed by Haegg (44). In this, a table of $\cos \theta$ for a set of standard values of θ is built into the instrument by means of a set of tappings on a transformer. The values of $h\xi$ are obtained by connections from the contacts on a ganged set of multiway switches, and the values of A_h are obtained by means of resistive potential dividers.

Another piece of equipment, designed by Story at Cambridge and at present under construction, works on a different principle, and evaluates the double sum (4.4) as a *continuous* function of x over the set of lines $y = 0.02(n+x)$ in the area $0 \leqslant x \leqslant 1$, $0 \leqslant y \leqslant 1$. A marker travels over a piece of paper in such a way that its position at any time represents the value of (x,y) for which the value of the sum is being carried out, and a mark is made on the paper whenever the value of the sum (4.3) becomes an integral multiple of some specified quantity. Thus the results are furnished in the form of a map of contours of constant value of the sum (4.4), which represents the distribution of electron density projected on the (x,y) plane.

Another instrument, devised by Pepinsky (92), uses electronic means of modulating the beam current in a cathode ray tube in such a way that the current to the point (x,y) in the plane of the screen is given by a sum of the form (4.4). This gives on the screen a direct visual display of the projection on the z-plane of the electron distribution in the crystal.

4.5. Integrating Instruments

Several instruments have been constructed for the evaluation of integrals, or the solution of differential equations of particular types, such as linear equations with constant coefficients. Such instruments have not been included in the chapter on the differential analyser, since one of the most important features of the differential analyser is its freedom from such a restriction to a particular type of differential equation.

(a) *Mechanical Integraphs.* A continuously variable gear, such as used in the differential analyser, is not the only kind of component that can be used as a mechanical integrator. Another is based on the use of a knife-edged wheel which, when moved over paper, cuts slightly into the paper and so is constrained to move in a direction given by the intersection of its plane with the plane of the paper. In a mechanical integraph, such a wheel is mounted so that it can turn about a vertical axis in its plane, and the mounting of this wheel is supported in a carriage which can move in the y-direction along a bridge which can move in the x-direction. If the plane of the wheel at any time makes an angle $\tan^{-1}z$ with the x axis, the path of the wheel is given by $dy/dx = z$. In the simple integraph (110), z is constrained by a linkwork to be proportional to the displacement w of an index, so that if the index is made to follow a curve $y = w(x)$, the wheel describes the curve $y = \int w(x)dx$. A graph of this integral curve is drawn by a pencil carried on the carriage which supports the mounting of the knife-edged wheel. By means of an adding linkwork, the slope dz/dx of the curve described by the wheel can be made to depend linearly on the displacements y of the wheel and w of the index, giving a means of obtaining graphically solutions of equations of the form

$$\frac{dy}{dx} = ay + bw.$$

Myers has developed further the integraph type of instrument, and has built one comprising two knife-edge integrating wheels and connecting linkwork, for obtaining graphical solution of linear differential equations of the second order with constant coefficients (86a).

(b) *Electrical Instruments.* Another type of instrument for evaluating solutions of linear differential equations with constant coefficients is one which depends on setting up an electrical circuit representing the equation to be studied. A particular example is provided by an electrical instrument, due originally to Beuken (10) and suggested independently by Paschkis (90, 91), for evaluating transient heat flow (see also ref. 71). If in the equation of heat conduction in one dimension

$$\frac{\partial\theta}{\partial t} = \frac{\partial^2\theta}{\partial x^2}$$

the space-interval is replaced by a finite difference (see §3.3), one obtains the set of simultaneous equations

$$\frac{d\theta_j}{dt} = \frac{\theta_{j+1} - 2\theta_j + \theta_{j-1}}{(\delta x)^2}$$

This set of equations is identical in form with the set of equations for the time variation of potential in a resistance-capacity ladder network (see fig. 35), so that, to the approximation represented by the replacement of the partial derivative by a finite difference, the temperature distribution in the heat flow problem can be obtained from a study of the potential distribution in the network, under initial and terminal conditions corresponding to those of the heat flow problem. This can be extended to heat flow in two or three dimensions (see fig. 36).

It may be possible to extend this method of treatment to heat flow in substances for which the thermal properties vary with the temperature,

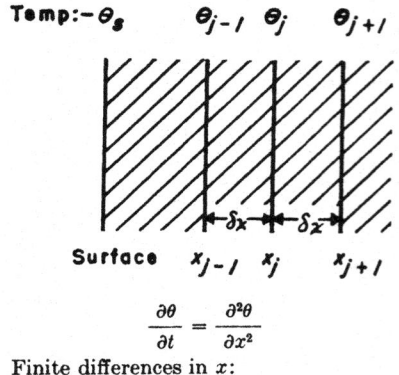

$$\frac{\partial \theta}{\partial t} = \frac{\partial^2 \theta}{\partial x^2}$$

Finite differences in x:

$$\frac{\partial \theta_i}{\partial t} \equiv \frac{\theta_{j+1} - 2\theta_j + \theta_{j-1}}{(\delta x)^2}$$

by changing the resistances (*not* the capacitances) appropriately in the course of a solution. Unless the thermal properties vary considerably, it would probably be best not to try to evaluate a solution in a single run, but to use an iterative process in each stage of which the thermal properties are input as given functions of time, obtained from the time variation of temperature obtained in the previous stage of the iteration.

If variations of terminal voltages to represent boundary temperatures, and variations of circuit parameters,

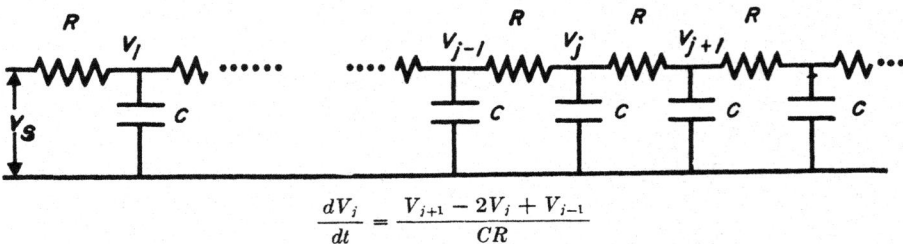

$$\frac{dV_i}{dt} = \frac{V_{i+1} - 2V_i + V_{i-1}}{CR}$$

FIG. 35. Electrical analogue for non-steady heat flow in one dimension.

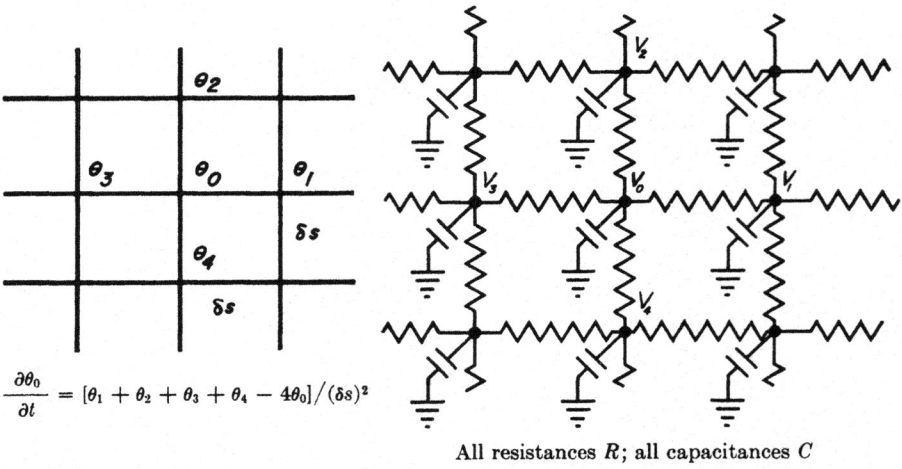

$$\frac{\partial \theta_0}{\partial t} = [\theta_1 + \theta_2 + \theta_3 + \theta_4 - 4\theta_0]/(\delta s)^2$$

All resistances R; all capacitances C

$$\frac{dV_0}{dt} = \frac{V_1 + V_2 + V_3 + V_4 - 4V_0}{CR}$$

FIG. 36. Electrical analogue for non-steady heat flow in two dimensions.

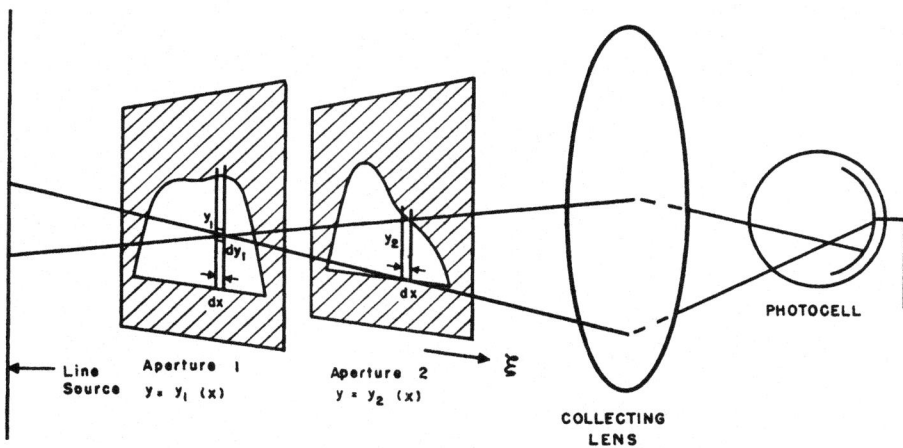

Fɪɢ. 37. Principle of the photo-electric integraph.

are made manually, the time constant of a loop of the circuit must be of the order of seconds, otherwise the operation of the instrument will be too fast for these variations to be made accurately enough. Thus the resistances must be of the order of megohms and the capacitances of microfacads, with correspondingly high leakage resistances.

(c) *Evaluation of Definite Integrals Which Are Functions of a Parameter.* For evaluating integrals of the type $\int_a^b f(x)g(x,\xi)dx$ as continuous functions of a parameter ξ, it is necessary to have a means of evaluating the integral for each value of ξ by an instantaneous process, as distinct from a method which builds up an integral from contributions evaluated during successive intervals over a period of time, as is done in the differential analyser, integraphs, or integrating circuits. A method of doing this is to measure the light flux through an aperture or set of apertures, which can in principle be done by the use of a collecting lens and a photocell.

An instrument of this kind is Gray's photoelectric integraph (43), later developed into an instrument called the cinema integraph (64). In this, light from a uniform line source parallel to the y-axis is passed through two apertures in different parallel planes, the x scales of the apertures being such that ordinates of the two apertures at the same value of x are coplanar with the source (see fig. 37). Then the light flux through an element of dy_1dx of one aperture and $y_2(x)dx$ of the other is $y_2(x)dy_1dx$, the total flux through $y_1(x)dx$ of one and $y_2(x)dx$ of the other is $y_1(x)y_2(x)dx$, and the total flux through the two apertures is $\int y_1(x)y_2(x)dx$. If all this light is collected by the photocell, and this is equally sensitive to light incident in any direction on the lens, the photocell current will be a measure of this integral. By displacing one aperture in the x direction, the integral $\int y_1(x) y_2(x - \xi) dx$ can be evaluated as a function of ξ.

Another instrument of this kind is one devised by Born and Fürth (13) for evaluating Fourier transforms of non-periodic functions. For this purpose it is necessary to evaluate $\int f(x) \cos kx \, dx$ and $\int f(x) \sin kx \, dx$ as functions of k. In this instrument a beam of light is formed with a sinusoidal variation of intensity over the wave front, and passed through an aperture cut to a shape determined by the curve $y = f(x)$ of the function whose Fourier transform is required. The light passing through this aperture is collected by a lens and focused on a photocell. The space-period of the sinusoidal variations of intensity can be altered continuously, and the voltage of one pair of plates of a cathode-ray tube is made to vary correspondingly. The output from the photocell controls the voltage on the other pair of plates, in such a way that a curve of the Fourier transform of $f(x)$ is displayed on the cathode-ray tube screen.

A difficulty with this method of carrying out integration by an instantaneous process is that negative values of the integrand cannot be handled as such. It is necessary either to determine separately the contributions from negative and positive values of the integrand, and to combine them at a later stage, or to bias the integrand so that it is always positive and later to subtract the additional contribution thus included in the integral actually evaluated. Either method is likely to lead to the final result being determined as a comparatively small difference of large contributions, with consequent limitation on its accuracy.

4.6. Directors

Another class of instruments, which should not go unmentioned but for which a bare mention must suffice, consists of the directors used for control of artillery fire against moving targets.

Consider the simplest case of a target moving on a horizontal plane through the gun, and approaching the gun with constant speed V. Let $T(\xi)$ be the time of flight of a shell to range ξ. If at any time the range to the target is x, and the gun is then fired at elevation for range ξ, the distance of the target at the time of fall of the shell is $x - VT(\xi)$, and for a hit this must be equal to the range ξ for which the gun was elevated (see fig. 38); that is

$$x - VT(\xi) = \xi$$

or

$$\xi + VT(\xi) = x.$$

This implicit equation must be solved continuously for ξ in terms of x, which is the quantity observed, and the function $T(\xi)$ occurring in it depends on the law of variation of resistance with velocity for the shell used, and is usually specified by a table or a graph rather than by a formula. Moreover, V can only be derived from the observed values of x,

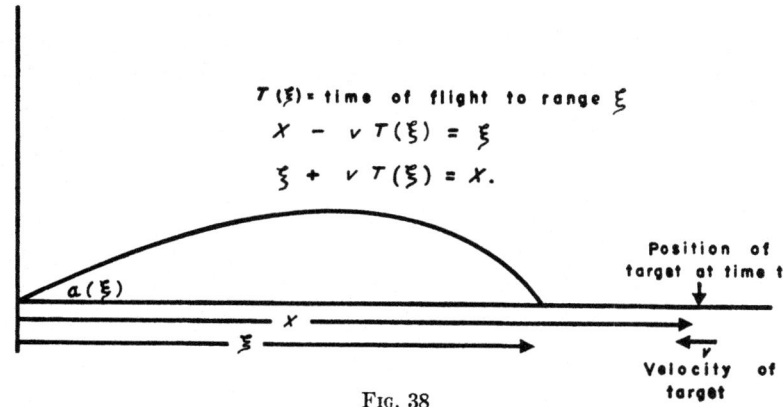

FIG. 38

all of which may be subject to errors of observation and measurement even if the actual target speed V is constant.

An actual director must be able to deal with the less simple case of a target moving at any angle to the line of sight, and further the equipment for the purpose, though of a precision character, must be robust and able to operate under active service conditions.

The development of such equipment is a considerable achievement, which would be more widely appreciated if the nature of its purpose did not put restrictions on the extent to which its design and performance could be made generally known.

Chapter 5

INTRODUCTION TO LARGE AUTOMATIC DIGITAL MACHINES

PRELIMINARY NOTE ON TERMINOLOGY

Any substantial scientific development involves building up a terminology, either of new words or of current words used in new or specialised senses, appropriate to the special features of the new development. There are disadvantages both in coining new words and in using current ones in special senses. Too free a use of new words gives the impression that the new development is something esoteric, closed except to those who have been initiated and learned the secret passwords, whereas specialised use of words already current may lead to misunderstanding, particularly when words habitually used in connection with living organisms, and especially with human activities, such as "memory," "choice," "judgment" are applied to mechanism.

Further, in the present case, different groups tend each to develop their own terminology, particularly appropriate to their own project, and an outsider trying to survey several projects has either to adopt the language of one group, and so appear to favour its particular ideas unduly, or to adopt, and to try to use consistently, a kind of average vocabulary, with perhaps some modifications of his own.

I have preferred to make use of current words rather than to invent new ones, and to use them in a way not specially related to that of any one group. A few terms need some comments.

'Instruction" is used, in preference to "order," for the statement of an operation the machine is required to carry out. This use of the longer word is preferred because "order" has another meaning which may be required in the same context, as when one speaks of "taking the operating instructions in order."

'Programming" and "coding" are apt to be used as almost synonymous. I have preferred to make a distinction between them, using "programming" for the process of drawing up the sequence of operations for a particular calculation, and "coding" for the process of translating either numbers or instructions into the form in which they are supplied to the machine, or in which they occur in the machine. In this sense, a number or an instruction can be "coded" on to an I.B.M. card or on to a set of relays. A calculation must be "programmed" in terms of a set of instructions which can be "coded" in accordance with the facilities of the machine.

'Control" seems to be used almost indiscriminately for the *operation* of taking the instructions in the appropriate order and initiating the appropriate action on each one, and for the *part of the machine* which carries out this operation. I have tried here to use 'control" for the operation, and "control unit" or "control system" for the hardware which carries it out.

I have preferred to use the neutral word "storage" or "store" rather than "memory," but have found no satisfactory neutral word to take the place of "judgment" as defined in §5.3; the alternatives "choice," "selection," "discrimination" which have been suggested seem unsatisfactory, and a much more comprehensive term is required to express the faculty involved here. For the selection of an instruction out of the normal sequence, on the basis of the assessment of one or more criteria, I have introduced the term "conditional selection" in place of the "conditional transfer" which is sometimes used, as without reference to a particular machine it is not obvious that any "transfer" is involved.

5.1. Historical

The main recent developments in the field of machines have been in the direction of large machines capable of carrying out, automatically, extended sequences of individual arithmetical operations, and so designed that the computing sequence can be changed from that required for one calculation to that required for another, so that the same machine can equally well be used, for example, for inverting a matrix, for the step-by-step numerical integration of a system of ordinary differential equations, or for the solution of multiple congruences.

The concept of such a machine is not by any means new. It is due to Charles Babbage, who was Lucasian Professor of Mathematics at the University of Cambridge from 1828 to 1839. But Babbage's conception of what he called the "analytical engine," like Kelvin's conception of a differential analyser, was ahead of its time; Kelvin's by about fifty years and Babbage's by about a hundred. Many of Babbage's ideas look remarkably fresh and modern when looked at in the light of recent developments, and a survey of them is given in Chapter 6.

The original idea of the analytical engine seems to have occurred to Babbage in 1835 or late 1834, and it was developed in the next few years. This was in the infancy of the science of electricity — the electrical relay was invented by Henry in 1835 — and Babbage's ideas for construction had to be in terms of purely mechanical components throughout. But many of the functions required in such a machine, particularly those concerned with control of the sequence of computing operations, and also those involved in simultaneous carry-over in addition, are much more easily carried out by electrical means, and these are freely used in all actual and prospective machines.

The first large automatic digital machine actually built was the I.B.M. Automatic Sequence-Controlled Calculator developed by Professor Howard Aiken and the I.B.M. Company, and installed at the Computation Laboratory at Harvard University (2, 62). This machine uses mechanical counters driven through electromagnetic clutches controlled by relay circuits. Although in construction, in its free use of electrical components, it is very different from Babbage's analytical engine, in principle, in what it does as distinct from how it does it, it is rather similar, and it has been hailed as "Babbage's dream come true" (23).

Other machines, one developed by the Bell Telephone Laboratories (120) and another by Professor Aiken (20a), use relays not only for control but for the storage and arithmetical units of the machines themselves.

Another, and the first to make use of electronic circuits, with the speed of operation of which they are capable, is the Eniac (39, 50), developed by Mauchly and Eckert for the Ballistics Research Laboratory at Aberdeen Proving Ground. More recently, another large machine has been completed by the I.B.M. Company (70).

These machines are considered in Chapter 7. Some later developments and projects are considered in Chapter 8. The remainder of this chapter is concerned with some general considerations relating to large digital machines.

5.2. Structure and Function in Calculating Machines

In the study of living organisms, a distinction is made between physiology, the study of the *functions* of the different parts of the organism in normal activity, anatomy, the study of their *structure*, and pathology, the study of the organism in disease. Similarly one can study a calculating machine — and other electrical and mechanical systems such as a railroad signalling system also — from the point of view of *structure* or *function*, or, if we take over the analogy from biology, from the point of view of their anatomy or their physiology. Machines also have their pathology, what happens when they go wrong, and diagnosis and cure of faults are important matters for the designer, who should see that these operations are easy, and for the maintenance staff; but this aspect need not concern us here.

A feature of a living organism is that the functions of its component parts are *co-ordinated*, and this co-ordination of functions is an essential feature of calculating machines also; it is this feature which makes the study of their "physiology" significant. The "anatomical" and "physiological" aspects are, of course, not independent, any more than they are for living organisms; but the distinction is a convenient one. The "anatomical" aspect of a calculating machine is all-important to the designer, and to the maintenance staff also, but the "physiological" aspect, the study of the machine from the point of view of function, is more valuable to the user, and to the non-specialist who wants to get a general idea of how the machine works without following out the mechanism or circuits in detail. I intend in these lectures to stress the "physiological" aspect, which may be the more general and apply to several machines, rather than the "anatomical" aspect which is peculiar to each machine.

5.3. Functions to Be Provided in an Automatic Digital Machine

Consider first the organisation of a calculation done by a human computer with the assistance of a desk calculating machine, as this will show what facilities have to be provided in order to carry out the same calculation automatically.

A human computer, carrying out a computation by pencil-and-paper methods with the assistance of a desk calculating machine, uses four kinds of equipment, the desk calculating machine for arithmetical operations, the working sheet for recording intermediate results and for keeping a note of the required sequence of computing operations, a set of volumes of tables, and his own mind for controlling the sequence of operations. The relations between these are indicated schematically in the block diagram fig. 39, in which the four rectangular blocks represent the four kinds of equipment, the lines indicate possible transfers between them (the line from the work sheet to the computer represents his use of the notes of the schedule of computing operations summarised on the work sheet), and the circles represent "gates" through which information can be transferred under the control of the computer.

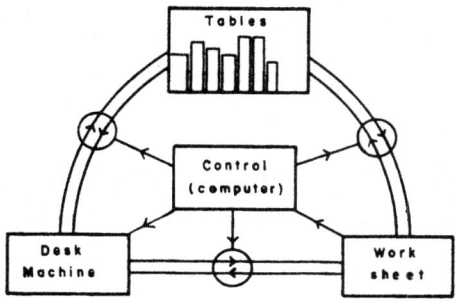

FIG. 39. Organisation of a
hand computation.

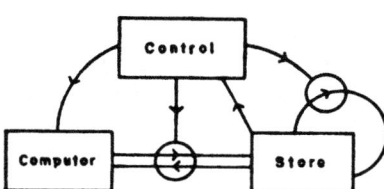

FIG. 40. Organisation of an automatic calculating machine.

An automatic machine can be thought of as having essentially the same functional organisation (see fig. 40). The functions to be provided in an automatic digital machine are:
  (i) Arithmetical operations [$+$, $(-)$, $\times . (\div, \vee)$, sign, $=$],
 (ii) Storage (for numbers and operating instructions),
(iii) Input (reception of data from the outside world),
 (iv) Output (supply of results to the outside world),
  (v) Transfer (of numbers or operating instructions from one part of the machine to another),
(vi) Control.
    Note. The components carrying out these different functions need not be all physically distinct. For example, a single unit may act both as a storage unit and as an adding unit.

It must of course have facilities for carrying out *arithmetical operations*, of which the essential ones are addition, multiplication, recognition of the sign of a number and of the equality of two numbers; subtraction can be replaced by the addition of a complement, and division, which may be rather an untidy process to mechanise in view of its trial-and-error nature, can be replaced by an iterative process involving only multiplication and subtraction; square root, which can be regarded as division by a variable divisor, can be treated in a similar way. These arithmetical operations are considered further in §5.5.

But facilities for carrying out arithmetical operations are by no means all that is wanted. The machine must have a *store* or "memory" both for numbers (data of the problem or intermediate results) and for the operating instructions. There must also be facilities for *input*, that is, for accepting data from the outside world, and this data must include any tabular functions required in the course of the work insofar as they are not evaluated by the machine itself as required; facilities are also required for *output*, that is, for furnishing results to the outside world, and for internal *transfer* of numbers between physically distinct parts of the machine.

Further, and most important, a control function must be provided by a unit taking the part of the computer in a hand calculation. This unit must be able to take the operating instructions in the appropriate order and ensure that the operations called for are carried out. In an extended calculation it often happens, not once but many times, that a stage is reached at which a departure from a regular repetition of a set of operations is required, or at which a selection has to be made between two, or perhaps more, alternative procedures for subsequent work. The control should have the faculty, for which the word "judgment" has been used (first by Babbage), of selecting the appropriate criterion to apply in such a situation, assessing the results of applying it, and on the basis of this assessment selecting the appropriate one of the alternative procedures.

Storage and control are aspects of a calculating machine at least as important as the means of carrying out arithmetical operations; the range of problems to which it can be applied effectively depends critically on the capacity of the store and the flexibility of the control. Further, the speed at which arithmetical operations can be carried out has a considerable influence on the form of the storage and control. With electronic equipment, arithmetical operations can be carried out in times of the order of milliseconds, and in order to make full use of this speed, the method of storage should be such that numbers can be recorded into the store and read out from it in times of the order of those required for the arithmetical operations, and similarly that the operating instructions should become available, and be acted on by the control unit, in times of a similar order of magnitude.

The arithmetical operations may be carried out by a structurally distinct portion of the machine, which can then be regarded as a unit, the "arithmetical unit" or "computor," but this is not necessary. This was to have been the case in Babbage's machine, but both in the Harvard Mark I machine and in the Eniac, addition is carried out in the storage units themselves, and there is no sharp structural division between "storage" units and "arithmetical" units.

5.4. Representation of Numbers in the Machine

In our ordinary digital representation of numbers, successive digits stand for coefficients of successive powers of ten, but there is no particular virtue, as far as a machine is concerned, in ten as a base; an alternative is to use a binary representation, in which the coefficients are all 0's and 1's and are coefficients of successive powers of two (fig. 41).

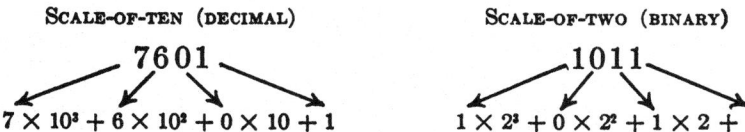

FIG. 41. Decimal and binary representation of numbers.

In some stages of work in a machine, or even throughout, it may be convenient to represent numbers by a set of elements each of which has only two distinct indications, for example by a set of relays each of which can be either "on" or "off," by a set of positions on a paper tape, at each of which a hole may or may not be punched, by a set of electronic tubes used as "on-off" elements, or by a set of signals each consisting of the presence or absence of an electrical pulse. The two indications can be regarded as representing "0" and "1," and use of such elements suggests the use of the binary representation of numbers in the machine.

On the other hand, use of such elements does not *necessarily* impose use of the binary representation; it is quite possible to code numbers expressed in decimal form on a set of such elements. This can be done in several ways. One way of coding decimal digits on such elements is to use a group of ten elements to represent each decimal place, and to arrange that one and only one has a "1" indication at any time; this method is used in the representation of numbers on I.B.M. cards, and in the Eniac. When the elements are pieces of hardware, such as relays or tubes, this probably leads to the simplest circuits but is extravagant in equipment by a factor of three; ten relays (for example) would be required to represent a single decimal digit, whereas, used for binary representation, they could represent all numbers from 0 to $2^{10} - 1 = 1023$, that is, three full decimal digits and a little more.

For short, a definite combination of indications of a set of elements will be called a *configuration* of the set. Since three such elements have only $2^3 = 8$ configurations, at least four are required to represent the ten distinct decimal digits. Four elements can be used in various ways; one is simply to use the binary representation of each decimal digit, but this has the disadvantage that it gives no positive indication of zero. Another, due to Stibitz, is to add three to each decimal digit and to represent the result in scale of two. This has three advantages: it gives a positive indication of the digit zero, complements are obtained by interchanging 0's and 1's, and in the process of addition carry-over from the most significant of the four binary digits occurs just when, and only when, carry-over from the corresponding decimal place is required. Other methods, using four, five, and seven elements to represent each decimal digit, have been proposed, and have certain advantages.

There is, of course, no need to base the design of a calculating machine on the use of elements giving only two distinct indications. This has to be done at any part of the machine in which relays or punched tapes are used, and with electronic circuits, design of the circuits and tolerances on characteristics of tubes and other circuit elements are easiest if the tubes are used as two-indication (on-off) elements, as is done in the Eniac and in almost all other electronic machines at present projected. But it is not in principle impossible to distinguish a number of discrete levels in the output from a tube, or in the magnitude of a pulse, and so to base a machine on the use, in part at least, of n-indication elements ($n > 2$). In this case, much of the attraction of the binary representation of numbers, which at present seems very marked, would disappear.

A machine working in terms of the decimal representation of numbers has one advantage which is not always appreciated by those who have not actually worked with a machine of this kind. In testing, in checking through the sequence of operating instructions, and in tracing faults either in the instructions or in the operation of the machine, it is most valuable to be able to carry out a set of instructions one by one and to compare the result at each stage with a value calculated in some other way. For example, in the step-by-step integration of an ordinary differential equation one might work through a representative interval of the integration with an ordinary desk machine to provide the comparison values.

The comparison is much more easily made if the results of the operation of the machine are displayed in scale of ten rather than in scale of two; it is much easier to check the agreement of two numbers each of ten digits 0 to 9 than two numbers each of 32 digits each of which can only be 0 or 1. In a binary machine, a binary-decimal conversion would be needed to provide a result in the required form, and this would not be satisfactory because if there were a disagreement, one could not know if the error were in the original result or in the conversion.

There are other ways of making the comparison, for example by making a decimal-binary conversion of the comparison figures and storing them in the machine, and making the comparison automatically. This would serve for routine testing, but not for tracing faults, since it will not always be known in advance at what point such a comparison is wanted, and it would load up the storage capacity too much to supply the machine with all the comparison data that might be wanted and all the operating instructions for using them.

These facilities of displaying results directly in decimal form, and of taking the operating instructions one by one, are both provided in the Eniac, and my experience was that both are most valuable, and form an important advantage of a decimal over a binary machine. It is rash to make guesses about future progress — or anyway, to proclaim them; but my own guess is that the use of the binary system in the machine is a passing phase, characteristic of the present stage of development. I am fully aware, though, that others whose opinions I respect dissent from this view.

5.5. Arithmetical Operations

(a) *Addition.* There are two ways of carrying out addition, firstly by counting, that is, by *successive* addition of *units*, and secondly by the use of an addition table. A machine working in scale of ten, and adding by

$b \backslash a$	0	1	2	3	4	5	6	7	8	9
0	0	1	2	3	4	5	6	7	8	9
1	1	2	3	4	5	6	7	8	9	10
2	2	3	4	5	6	7	8	9	10	11
3	3	4	5	6	7	8	9	10	11	12
4	4	5	6	7	8	9	10	11	12	13
5	5	6	7	8	9	10	11	12	13	14
6	6	7	8	9	10	11	12	13	14	15
7	7	8	9	10	11	12	13	14	15	16
8	8	9	10	11	12	13	14	15	16	17
9	9	10	11	12	13	14	15	16	17	18

FIG. 42. Addition table.

counting, has to be provided with means for carrying out the following arithmetical operations:

$$a + 0 = a$$

$$0 + 1 = 1 \qquad 5 + 1 = 6$$
$$1 + 1 = 2 \qquad 6 + 1 = 7$$
$$2 + 1 = 3 \qquad 7 + 1 = 8$$
$$3 + 1 = 4 \qquad 8 + 1 = 9$$
$$4 + 1 = 5 \qquad 9 + 1 = 10$$

whereas one working in scale of ten and adding by means of an addition table has to be provided with means for carrying out the whole set of operations represented by fig. 42. In scale of two, there is no distinction between the two methods; each involves just

$$0 + 0 = 0, \quad 0 + 1 = 1 + 0 = 1, \quad 1 + 1 = 10.$$

Further, addition in the different decimal (or binary) places can be carried out successively or simultaneously (see fig. 43). In the latter case, the addition is carried out in two stages; in the first stage, addition is carried out in each place separately, the carry-over digits being recorded to be dealt with at the second stage, and the carry-over can also be made successively or simultaneously. It might seem at first sight that carry-over would have to be successive, starting with the least significant place, since until the carry-over *into* one place has been made, it is not known whether there will be carry-over *from* it to the next place. However, as Babbage knew, simultaneous carry-over is possible. The method is as follows.

ADDITION

```
successive                 simultaneous addition
addition                          successive        simultaneous
                                   carry               carry
 6374          6374              398463              398463
 2138          2138              101718              101718
 ────          ────              ──────              ──────
    2          8402  ⎫ first    ⎰499171             499171
 1    carry      11  carry⎬process⎱  1 1   carry     001 1   carry
 ────          ────       second     1              ──────
 1             8512       process    0   c.         500181
 1    carry    ════                  8
 ────                                0   c.
 5                                   1
 0    carry                          0   c.
 ────                                0
 8                                   1   c.
 ════                                0
 8512 Ans.                           1   c.
                                     5
                                     ──────
                                     500181 Ans.
```

FIG. 43

The sum of two single decimal digits is a number of two digits of which the left-hand (more significant) is 0 or 1 and the right-hand may be any digit from 0 to 9; but *if the right-hand digit is 9, the left-hand digit must be 0*. If the right-hand digit is not 9, no carry-over into this place will result in carry-over from it; if the right-hand digit is 9, the left-hand is 0, and there is no carry-over *from* this place unless there is a carry-over *into* it, and then the result in this place is 0. Hence a carry-over is only propagated through a number of places if as a result of the first stage there is a series of 9's; then a carry-over into the least significant of the 9's turns all the 9's to 0's and gives a carry-over of 1 from the most significant of them; otherwise these digits and the next more significant are unaffected. The argument holds in any other scale of notation, to base n, if $(n-1)$ is substituted for 9.

A method of addition with simultaneous carry-over, based on this, was to have been used in Babbage's analytical engine, and is used in the Eniac (both these in scale of ten) and in Booth's Automatic Relay Computer (scale of two).

(b) *Multiplication.* Multiplication can be carried out by successive addition, by the use of a built-in multiplication table (as in the Eniac), or, in a decimal machine, by building up the nine multiples of the multiplicand, selecting the appropriate multiples according to the digits of the multiplier, and adding them with appropriate shifts (this method is used in the Harvard Mark I Calculator).

(c) *Division and Square Root.* Reciprocals can be found by an iterative method which involves only multiplication and subtraction, and then x/a can be found by multiplying x by $(1/a)$. The iterative method is as follows. Let y_n be a sequence defined by

$$y_{n+1} = y_n (2 - ay_n).$$

Then, provided y_0 lies within certain limits, $y_n \to 1/a$. This iterative process is "second order" (see Chapter 9). For a machine working in scale of ten, it may be convenient to bring any number into the range $2 \leqslant z < 5$ by multiplication by a suitable power of ten and by 2 or 5, factors which can easily be replaced when the reciprocal is found; then by taking $y_0 = 2/7$, one is assured of 9-place accuracy in y_n after 5 iterations.

Division can also be carried out by a process similar to that used for automatic division in desk machines (as in the Eniac) or by building up the series of multiples, from 1 to 9 times, of the divisor and at each stage comparing the remainder with them and subtracting the greatest which will still leave a positive remainder. It should be mentioned that there are two forms in which an answer to a division may be required; one is the form of an integral quotient and a remainder; the other is a quotient in decimal (or binary) form to a certain number of significant figures or decimal places. The use of the iterative method for a reciprocal is only useful if the second form of result is the one required.

An iterative method can be used for square root. One method is to use Newton's formula

$$y_{n+1} = \tfrac{1}{2} \left[y_n + (a/y_n) \right] \tag{5.1}$$

$(y \to a^{\frac{1}{2}})$, but since this involves a division, it may not be very convenient. Another iterative formula for a square root is

$$y_{n+1} = y_n \left[3a - y_n^2 \right]/2a, \tag{5.2}$$

which has the advantage that only one division is required, namely a preliminary calculation of $1/2a$, whereas formula (5.1) requires a new division for each repetition of the iterative process.

(d) *Recognition of the Equality of Two Numbers.* Recognition of the equality of two numbers can be derived from the operation of recognition of the sign of a number, for if both $(x - y)$ and $(y - x)$ are non-negative, x must be equal to y. This is the method that has to be used on the Eniac. However, the equality of two numbers is so often a useful criterion to use as a basis for selecting between alternative procedures that it seems advisable to include the recognition of equality (or of inequality) as a fundamental operation.

(e) *Signs and Decimal (or Binary) Points.* Other features of a calculating machine are how it deals with signs and with the decimal or binary point.

There are two ways of indicating negative numbers, by a sign indication and either the modulus or the complement of the number. Both methods are represented in the machines now in operation or in course of development. If a sign indication and modulus are used, addition and subtraction can be carried out by a unit which can evaluate $|a| + |b|$ and $||a| - |b||$, selection between these processes in any particular case being made according to the signs of a and b and the operation to be performed on them; the appropriate sign indication is then appended to the result. In multiplication and division the signs and moduli can be treated separately.

There are three ways of treating the position of the decimal (or binary) point. Each of these has its disadvantages, and probably more practical experience than is at present available is necessary before it can be said definitely which is the preferable.

One ("floating decimal* point") is to specify numbers by a number between certain limits (between 1 and 10, or between 0.1 and 1, these being read in whatever scale of notation is being used) and the index of an integral power of ten (or two). Another ("fixed decimal point") is to regard the position of the decimal point in all numbers in the machine as being fixed, so that these numbers are all between the limits (say) ± 1; this involves the use of scale factors for representing in the machine numbers which are greater than, or very much less than, 1; if the orders of magnitude are approximately known beforehand, this can be done without numbers either becoming too large for the machine or becoming so small that the number of *significant* figures retained is inadequate; but if orders of magnitude are not known beforehand, either of these difficulties may arise, or alternatively, an elaborate series of conditional instructions is necessary to ensure that they do not.

The third ("programmed decimal point") is to regard the position of the decimal point as being determined by the operating instructions and the programming, rather than by the form of the numbers in the machine. This imposes on the operator some of the burden which it would be reasonable to expect the machine to bear.

---

* For "decimal" read "binary" throughout if numbers are expressed in scale of two.

A floating decimal point complicates the process of addition, since the indices of the powers of ten (or two) in the numbers to be added have to be compared and the appropriate shifts made, as a preliminary to every addition. This is a rather serious objection, since addition (including subtraction) is the commonest single operation of all, and one which it is most desirable to keep simple and rapid. On the other hand, for multiplication a floating decimal point is very satisfactory and means that the full capacity of the machine is always used. However, there are occasions on which one does not want to be obliged to have numbers calculated in the form $a \times 10^n$ $(0 \leqslant |a| \leqslant 1)$; for example if z has been calculated and y is being calculated solely for the purpose of being added to z, it is most convenient to have the decimal point in the same place for y as for z. This is a common situation; for example z may be an indefinite integral and y the contribution to it for an interval of the independent variable. Further, in this case one does not want the position of the decimal point in z in the machine to be shifted as z passes through zero. It would rather seem that some combination of the use of a floating and a fixed or programmed decimal point would be the most useful.

5.6. Serial (Successive) and Parallel (Simultaneous) Operations

Reference has already been made in §5.5(a) to a distinction between two ways of carrying out addition, according as the different decimal (or binary) places are treated successively or simultaneously. This is an example of a wider distinction between what is often called "serial" and "parallel" operation.

The representation of numbers themselves can be made in serial or parallel form. For example, if the digits of a number expressed in scale of two are represented by the presence or absence of electrical pulses (indicating "1" and "0" respectively), a 30-place number could be represented either by a *succession* in time of "0" or "1" indications on a single line (serial representation) or by a set of *simultaneous* "0" or "1" indications on thirty lines (parallel representation).

Serial representation of numbers leads naturally to serial treatment of the binary (or decimal) places within a single arithmetical operation. If this form of representation is used, it is a consequence of the process of carry-over in addition that the *least* significant digit of a number should come first in time, contrary to our usual way of reading numbers; this corresponds to the ordinary procedure in arithmetic, of "adding from right to left." Similarly a parallel representation of numbers leads naturally to a parallel (simultaneous) treatment of the different binary (or decimal) places, including simultaneous carry-over.

Serial operation has the advantage that the only equipment needed for addition is a unit for carrying out addition in a *single* binary (or decimal) place; the same piece of hardware deals with each binary (or decimal) place in succession, whereas for parallel operation the adding unit has to be duplicated for each binary (or decimal) place in the numbers to be added. Correspondingly, serial operation can easily be applied to numbers of any length, whereas parallel operation is restricted to numbers of lengths within the capacity of the number of adding units.

On the other hand, for a given pulse frequency, parallel operation is substantially faster; or, to obtain a given over-all operating speed, a higher pulse frequency has to be used for series than for parallel operation.

'Serial" and "parallel" operation may also refer to arithmetical and other operations considered as units. Multiplication, in most machines, takes a long time compared to addition, and time could be saved by carrying out additions and other operations while a multiplication is in progress, if the organization of the machine and the sequence of computing operations required were such as to make this possible. This is called "parallel operation" as distinct from "series operation" in which only one arithmetical operation can be carried out at once.

5.7. Static and Dynamic Storage

A number may be represented in the store, or elsewhere in the machine, by a configuration of counting wheels, relays, or tubes; such a configuration is static and can be read whenever required. Such a form of storage is called "static storage."

A number may also be represented by a succession of signals travelling along a delay line, or impressed on magnetic material in the form of surface of a drum, which rotates under one or more reading heads. These signals are moving relatively to the places at which they can be read, and only become available at certain times, when the signals in the delay line reach the end of the line, or when those in the magnetic material pass under a reading head. Such a form of storage is called "dynamic storage."

Use of dynamic storage involves timing arrangements, in the operating instructions and in the design of the machine, to record at what time each number in the store becomes available and to ensure that any steps taken to read it out from the store are taken just as it becomes available. On the other hand, with various forms of dynamic storage, considerable capacity can be achieved without a great deal of equipment (see §8.2).

Static storage avoids most of the timing problems involved in dynamic storage, but, at present, involves a large amount of equipment. Storage of 1000 numbers, of 10 decimal digits each, on relays, or on tubes used as on-off elements, would require over 30,000 relays or tubes for number-storage alone; with dynamic storage using delay lines, something like 1000 tubes would be required. The development of a reliable and satisfactory form of static storage, into which numbers could be recorded and from which they would be read in times of the order of milliseconds, and economical of electronic equipment, would make static storage a practical alternative where now, in view of the capacity required, only some form of dynamic storage is practicable.

A convenient way of using dynamic storage is to record the binary digits of a number successively in one storage element (a delay line or magnetic tape, for example) so that they are read out serially and are in a form for immediate use in serial methods of carrying out addition, etc.; hence dynamic storage tends to be associated with serial operation, and similarly static storage with parallel operation (as in Babbage's analytical engine). But this association is not necessary; another way of using dynamic storage is to record one digit of a 30-digit binary number (for example) on each of 30 storage elements, so synchronized that all the digits of any one number become available together, and can be used in a parallel method of carrying out addition (as is done in Booth's Automatic Relay Computer). On the other hand, the contents of a static storage system could be scanned in sequence so as to produce a serial representation of a number contained in the store.

5.8. Control and the Form of the Operating Instructions

It has already been mentioned that storage and control are aspects of a calculating machine at least as important as the means of carrying out arithmetical operations. And these aspects are related, since the organisation of the control unit will depend on the form in which the operating instructions are furnished to it, and on the form in which the numbers to be operated on are stored. For example, with a dynamic form of storage the control unit must include some kind of clock to determine the time at which the required information becomes available, whereas with a static form of storage questions of timing are not so prominent; those which do arise do not continually involve such precise timing as this, but are mainly concerned with ensuring that one operation is complete before another is started.

The organisation of the control system depends on the character of the form of instruction adopted as standard. What this means can be seen by considering a particular example. Suppose it is required to add the numbers in locations A and B in the store and to place the result in location C. In a machine with a distinct arithmetical unit containing an accumulator to or from which numbers can be transferred directly, this could be done by the following three separate instructions:

(i) Clear accumulator and transfer number in location A to it,
(ii) Transfer number in location B to accumulator,   (5.3)
(iii) Transfer number in accumulator to location C.

A single instruction of this type specifies two locations, one in the store and one in the arithmetical unit, and an operation (transfer being regarded as an operation). An alternative is to use instructions of such a form that the whole process is covered by a single instruction:

>   (i) Form the sum of the numbers in locations A and B $\left.\vphantom{\begin{matrix}1\\1\end{matrix}}\right\}$ (5.4)
>   and place the result in location C.

A single instruction of this form specifies three locations in the store and an operation.

Another relation between the organisation of the control system and the standard form of instruction adopted is involved in the way in which the source of the next instruction is determined. Here there are two situations to be met. Consider a stage in which the operation specified by one instruction, which will be called the "current instruction," is being carried out. The next instruction may either be specified unconditionally or it may have to be selected according to a criterion, evaluated by the operation specified by the current instruction or earlier. The latter process will be called "conditional selection" (it is sometimes called "conditional transfer"; but see the preliminary note on nomenclature at the beginning of this chapter). A sequence of operation in which no conditional selection is involved will be called a "linear sequence."

For any instruction in a linear sequence, the location of the next instruction can be specified in the current instruction, but even this may not be necessary. If these instructions are stored serially in the order in which they occur in a linear sequence, then a counter, so connected that its content increases by unity for each instruction operated on, will automatically specify the location of the next instruction.

It has become customary to specify the form of instruction by using the term "n-address code" for a way of coding instructions in which each instruction refers to n storage locations. For example, a method of coding which uses instructions of the form (5.3), stored in serial order so that no explicit statement of the location of the next instruction is required, is referred to as forming a "one-address code," whereas a method using instructions of the form (5.4) with the addition of an explicit statement of the location of the next instruction, is called a "four-address code." This terminology is more appropriate to some machines than to others.

A single instruction of a one-address code is simpler than a single instruction of a three- or four-address code, and can probably be arranged to take half the space in the storage system or less. But the number of instructions is considerably larger in the former case, and probably the total storage capacity would usually be somewhat larger. On the other hand, programming and coding can probably be made easier with the larger number of simpler instructions.

Chapter 6

CHARLES BABBAGE AND THE ANALYTICAL ENGINE

6.1. Babbage's Calculating Engines

The general idea of a large, general-purpose, automatic calculating machine is due to Charles Babbage, who was Lucasian Professor of Mathematics at the University of Cambridge from 1828 to 1839. Babbage was concerned with two different kinds of calculating machine, which he called respectively the "difference engine" and the "analytical engine." The purpose of the difference engine was to build up tables of functions from high-order finite differences; this is the better-known, as it was partly constructed and portions can be seen in various scientific museums, etc. But much the more interesting in view of recent developments is the machine which he called the "analytical engine," which was to be very much more versatile than the difference engine; indeed it was to be a large, almost automatic, general-purpose calculating machine.

Babbage wrote a book of comments on his work and experiences, rather too discursive to be called an autobiography, entitled "Passages from the Life of a Philosopher."* Some chapters of this book are devoted to the difference engine and the analytical engine, but the accounts, both of the general ideas and the proposals for design and construction of hardware, are not at all detailed or systematic. Babbage, however, visited Italy about 1840-41 to discuss his ideas for the analytical engine with interested people in that country, and a General Menebrea afterwards wrote up his notes of the discussions in a paper (81) published in the Bibliothèque Universelle de Genève. This paper was translated by the Countess of Lovelace, the daughter of the poet Byron, and she increased its value very much by adding translator's notes to the extent of more than twice the original paper, with fuller analytical discussion of various points, and examples of what would nowadays be called "programming" of problems for the analytical engine. This translation with notes was published in Taylor's "Scientific Memoirs" (82). It is included in a collection of papers published by Charles Babbage's son, H. P. Babbage, entitled "Babbage's Calculating Engines,"† and is the most detailed source of information on Charles Babbage's ideas on this subject that I know. A draft, in the same volume (C, p. 330), of a paper by H. P. Babbage to the British Association in 1888 gives the fullest account of the design of the hardware. (In the British Association Report for this year, there is only a brief summary of this paper.)

It is not clear how Lady Lovelace came to be interested in the analytical engine or to have such a clear and detailed appreciation of the

---

* Ref. 6. Here referred to, for brevity, as P.
† Ref. 7. Here referred to as C.

project as her notes show; it must be remembered that nothing of this machine had been constructed at the time. She must have been a mathematician of some ability. Babbage (P, p. 136) quotes her as having corrected his analysis in the derivation of a recurrence formula for the Bernoulli numbers which is used as an example of programming, and she discusses at some length the structure of extended numerical calculations and the way in which any repetitive features can be exploited in the programming, and shows appreciation of the way in which computing problems have to be analysed to make them accessible to the machine.

Some of her comments sound remarkably modern. One is very appropriate to a discussion there was in England which arose from a tendency, even in the more responsible press, to use the term "electronic brain" for equipment such as electronic calculating machines, automatic pilots for aircraft, etc. I considered it necessary to protest against this usage (51), as the term would suggest to the layman that equipment of this kind could "think for itself," whereas this is just what it cannot do; all the thinking has to be done beforehand by the designer and by the operator who provides the operating instructions for the particular problem; all the machine can do is to follow these instructions exactly, and this is true even though they involve the faculty of "judgment." I found afterwards that over a hundred years ago Lady Lovelace had put the point firmly and concisely (C, p. 44): "The Analytical Engine has no pretensions whatever to *originate* anything. It can do whatever *we know how to order it* to perform" (her italics).

This does not imply that it may not be possible to construct electronic equipment which will "think for itself," or in which, in biological terms, one could set up a conditioned reflex, which would serve as a basis for "learning." Whether this is possible in principle or not is a stimulating and exciting question suggested by some of these recent developments (see Ch. 8). But it did not seem that the machines constructed or projected at the time had this property.

6.2. Babbage's Analytical Engine

The analytical engine was to consist of three main parts: one, which Babbage called a "store," in which numerical information could be recorded on a bank of counters; another, which he called the "mill," in which numerical operations could be carried out on numbers taken from the store; and a unit to which he did not give a name, but which we may call the "control unit," which should control the sequence of operations, the selection of the numbers on which they were to be performed, and the disposal of the result.

Some parts of this machine appear to have been made in Babbage's lifetime. After his death, part of the "mill" was built by his son, H. P.

Babbage, and is now in the Science Museum in London. This machine was to work in scale of ten, and to have 50-figure capacity. Addition was to be simultaneous in all decimal places, and Babbage saw that if carry-over was to be successive over 50 places it would take a long time compared with the first and main stage of the addition process (see §5.5(a)). He seems to have spent much time on the problem of simultaneous carry-over, first to find in principle how it could be done and then to devise a means of doing it mechanically in practice. He was successful, and a 29-place adding mechanism with simultaneous carry-over was later constructed by H. P. Babbage, and reported to work satisfactorily (C, p. 335).

In view of recent developments, it is interesting to see what capacity Babbage planned for his "store"; this was to be 1000 numbers of 50 decimal digits. He estimated that its speed would be

> 60 additions per minute
> 1 multiplication of two numbers of 50 digits per minute
> 1 division of a number of 100 digits by one of 50 digits per minute.

In one respect, the analytical engine was not automatic. Tabular functions were to be punched on cards ("number cards"); if the value of such a function was wanted, the machine was to display the argument value for which it wanted the function value, and to ring a bell. An attendant was then to pick out the card with the specified argument value, and insert it into the machine, which would verify that it had been supplied with the right card, and would read off the function value. If by mistake it had been supplied with the wrong card, it would "ring a louder bell, and stop itself" and display a notice "wrong number card." Presumably there would also be a check that the card referred to the right function, though this is not mentioned. Babbage planned that number cards representing standard functions would themselves be calculated and punched by the machine, and would therefore be certain to be correct.

6.3. Control in the Analytical Engine

Control of the sequence of computing processes was to be carried out through a set of punched cards, like the cards used in a Jacquard loom for the mechanical weaving of damasks and other elaborately patterned fabrics. Plungers passing through holes in the cards were to operate the mechanism for selecting the counters in the "store" from which numbers were to be transferred to the "mill," the arithmetical operation to be performed on them, and the counter to which the result was to be transferred. There were to be two sets of such cards, "operation cards" for specifying the sequence of operations, and "variable cards" for specifying the selection of counters for these transfers.

Normally the cards would be taken in sequence, but there were to be means of advancing or backing the cards to any specified extent if required, according to the result of some criterion or set of criteria previously evaluated in the course of the calculation. Normally, the control of this conditional selection of the next instruction to be satisfied was to be made by means of an indication of carry-over from the most significant digit of a counter, which would occur or not according to the sign of the difference of two numbers. By this means the evaluation of one or more such criteria could be used as a basis for selection between different alternative procedures (P, p. 135).

In several accounts of Babbage's analytical engine there is no mention of this faculty of "judgment," and it almost seems as if its importance had not been realised until quite recently. But Babbage himself was clearly aware of this, and had seen how to provide this faculty by mechanical means. It is doubtful, though, if even he realised the full significance of it. In its effect on the range and versatility of the machine it seems now of much greater importance than simultaneous carry-over in addition, which Babbage explicitly says he regards as the most important feature of the analytical engine (P, p. 114).

6.4. Organization of Calculations for the Analytical Engine

The separation of the cards controlling the operations of the analytical engine into two groups, "variable cards" specifying locations in the store from which numbers were to be transferred to the mill and to which results were to be transferred from the mill, and "operation cards" specifying what operations were to be carried out in the mill, led to a way of looking at the structure of a calculation in terms primarily of the sequence of operations involved, independent of the numbers on which these operations were carried out.

This led to the recognition of the repetitive nature of many extended calculations, and the ideas of a recurring group of operations, which is termed a "cycle," and of a recurring group of such cycles, or "cycle of cycles" (C, p. 39). Further, it is recognised that in a recurring group of cycles which consists simply of the repetition n times of a single cycle, the number n may vary from one recurrence of the group of cycles to another, and that provision must be made for this. It is also recognised what an economy of operation cards would be effected by programming the recurring cycles of operations just once each and using other facilities of the control system to organize the repetitions of these cycles; the solution of simultaneous linear algebraic equations by elimination is taken as an example (C, p. 42).

These ideas are not prominent in Babbage's own writings, but they are developed in Lady Lovelace's notes to the translation of Menebrea's paper, and a notation is developed to treat this aspect of the organization of repetitive calculations. As an example, a linear recurrence relation for the Bernoulli numbers is obtained, giving B_{2n} as a linear combination of Bernoulli numbers of all lower order, with coefficients which are functions of n; thus both the numbers of terms in the recurrence relation and the values of their coefficients vary with n. The evaluation of the Bernoulli numbers from this recurrence relation is programmed for the analytical engine.

Chapter 7

THE FIRST STAGE OF DEVELOPMENT

7.1. The Harvard Mark I Calculator

The Harvard Mark I Calculator was planned before the war, but only completed after the entry of the United States into the war. A full account of it, and of the process of programming and coding problems for it, has been published by the Harvard University Press (62); here this will be referred to briefly as M. The present account is only a summary of the main features of this machine, from the functional rather than from the structural point of view (see §5.2).

In principle and in its functions, it is broadly similar to Babbage's conception of the Analytical Engine, and is the first machine actually to be constructed and operating of which this can be said, though with its very extensive use of electrical components, it is structurally very different from anything Babbage could have foreseen. As the Analytical Engine was to do, it works directly with numbers in their decimal form, and has a static form of storage. In some of its functions it differs from the Analytical Engine; for example, the process of referring to tables of functions is automatic, once the machine has been set up for a particular problem and supplied with the required information, and does not need the attention of an operator. Also there is no sharp distinction between the storage system and the portion of the machine which carried out the arithmetical operations, as there was to be between the "store" and the "mill" of the Analytical Engine; the counters which form part of the "store" are also adding units.

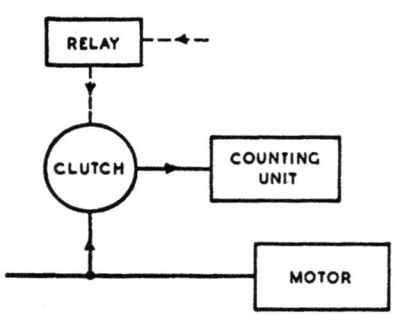

Fig. 44. Basic principle of the automatic sequence-controlled calculator.
————— mechanical connections
- - - - - - - - electrical connections

The basic principle of those parts of the machine which carry out arithmetical operations is represented very schematically in fig. 44. A shaft is driven continuously by a motor, and rotation can be transferred from it to mechanical counting units through electromagnetic clutches controlled by relays, which may themselves be controlled by circuits made through contacts on other relays or on other counting units, through switches, or through holes in punched cards. A counting unit has ten discrete positions, and the shaft, though continuously rotating, can be regarded as a source of mechanical impulses each of which can move a counting unit from one of its positions to the next. A counting unit also carries a set of contacts, one for each of its ten positions, and other contacts involved in the process of carry-over in addition. There is one such counting unit for each of the 23 decimal places of each of the 72 counters which form the main storage of the machine for intermediate results; each counter also has a similar unit for sign indication. Other counters are associated with the multiply-divide unit and other parts of the machine.

Storage for data of the problem, and for constants whose values are known before the work starts, is provided by a set of ten-position hand switches, on which 23-figure values of 60 such quantities, with sign indications, can be set by hand. Storage for tabular functions is provided by paper tapes, on which information for interpolation (for example, finite differences or reduced derivatives, according to the interpolation process to be followed) as well as function values can be punched. Numerical data on such tapes is represented in coded decimal form, using a four-hole code.

Addition is carried out by counting (i.e., by successive addition of units, see §5.5(a)), and the counters are used for this purpose as well as for storage units; it is only for multiplication and division that numbers have to be transferred from the counters to a separate unit for the arithmetical operations. This multiply-divide unit builds up the set of multiples, from 1 to 9 times, of the multiplicand or divisor. Then in multiplication it selects the appropriate multiples, according to the digits of the multiplier, and adds them with appropriate shifts. In division the remainder at any stage is compared with this set of multiples, and the largest which can be subtracted and leave a positive remainder is subtracted from it; the remainder after this process is then shifted and the process repeated. Before starting a division, both dividend and divisor are shifted, if necessary, so that the most significant figure occurs in the furthest left-hand place in the corresponding register of the unit; a record is made of the difference between these shifts, and the position of the decimal point in the quotient is adjusted accordingly. The programming of the operation involved in a multiplication or division is built into the multiply-divide unit, so that in coding a particular problem, the instructions for providing the unit with multiplier and multiplicand, or divisor and dividend, and for specifying the operation to be performed on them, are all that have to be coded.

The machine is also furnished with means of evaluating logarithms and antilogarithms to base ten, and sines. A table of 36 logarithms, of the numbers 1(1)9, 1.1(0.1)1.9, 1.01(0.01)1.09, 1.001(0.001)1.009 is permanently built into the "logarithm unit" of the machine. By a series of divisions, a number Z whose logarithm is required is expressed in the form $Z = 10^n\,abcdz$, where n is an integer, a, $10(b-1)$, $100(c-1)$, $1000(d-1)$

Fig. 45. General front view of automatic sequence-controlled calculator
(Harvard Mark I Calculator).

are all integers each consisting of a single digit, and $1 \leqslant z < 1.001$. Then $\log a$, $\log b$, $\log c$, and $\log d$ are all included in the log table built into the machine; $\log z$ is evaluated from the power series, of which the coefficients are also built into the machine. Antilogarithms are calculated by a process which is effectively the inverse of this. The programming of the required sequence of operations is also built in, so that each of these whole sequences is coded as a single instruction.

For the evaluation of the sine of an angle, the argument for which the sine is required is first brought into the range $0 \leqslant \theta \leqslant \pi/2$, and the sine is then calculated from the power series, of which the coefficients are built into the machine. As in the calculation of logarithms, the programming of the sequence of operations required in the calculation of a sine is also built in, so that only a single instruction is required in programming any particular problem.

An alternative method for calculating sines would be to follow a process similar to that used for logarithms and antilogarithms, by building in, say, a 19-value table of $\sin \theta$ for

$$\theta/\tfrac{1}{2}\pi = 0.1(0.1)1.0, \ 0.01(0.01)0.09$$

from which the sine and cosine of any angle which is an integral multiple of $(0.01)\pi/2$ can easily be built up, leaving only the sine and cosine of an angle of magnitude less than $(0.01)\pi/2$ to be evaluated from the series.

Other functions can be supplied to the machine in coded form by punched tapes ("function tapes"), as already mentioned. The tape is fed through the reading head of an interpolator, which first searches for the required argument value and then reads the function value and interpolation data and controls the sequence of operations for carrying out any interpolation required. There are three independent interpolators, so that three distinct functions can be supplied to the machine if required.

Results are output by means of two electrical typewriters, or on standard I.B.M. cards by means of a card punch. Data can also be supplied to the machine from punched cards. These have to be used in the order in which they are stacked in the deck, and can only be used once each (unless manipulated by hand in the course of the calculation), so they do not form a suitable means of inputting tabular functions, other than functions of an independent variable for which the required values and the order in which they will be required are known before the calculation is started. Use of cards, however, is valuable in extending the storage capacity of the machine for intermediate results beyond that provided by the 72 counters.

Figure 45 shows a general front view of the machine. On the extreme left of fig. 45 are the switches for setting values of constants and other data; to the right of these are the banks of counters, and beyond these

Fig. 46. Sequence control mechanism of Harvard Mark II Calculator.

the multiply-divide unit and the units for calculating logarithms, anti-logarithms, and sines. To the right of these are the units for reading the function tapes and the sequence control tape, each on a panel with associated switches, and then on the extreme right the output units, two electrical typewriters above and a card punch below, where there are also two card readers. The control relays are mainly in racks behind.

7.2. Control in the Harvard Mark I Machine

The sequence of operations is specified to the machine in coded form by means of punchings on a paper tape. Holes can be punched in any of 24 positions in a line across the tape, and successive lines correspond to successive instructions. This tape, the sequence control tape, is passed through a reading mechanism in which electrical contacts are made through the punched holes. Figure 46 shows the reading mechanism for the sequence control unit, with a sequence control tape in place. The 24 positions for the holes are divided into three groups of eight, of which two groups specify the source and destination for the transfer of a number, and the third specifies an operation to be then carried out, so that the standard verbal form of an instruction is "Read the number in place A; transfer it to place B; start operation C." When the content of a counter is read out from it, it is also held in the counter; an explicit instruction is required to clear it.

In the machine as originally built, the operation of the sequence control tape was as follows. On completion of the operation specified by one row of holes, the sequence control mechanism steps the tape on to the next row. Thus if an iterative process forms part of a longer computing sequence, each repetition of the iteration has to be coded separately, and (apart from manual interference with the sequence of computing operations) there is no provision for controlling the number of repetitions of the iteration according to the results of the successive repetitions. The only automatic selection between operating instructions is a selection between taking the next instruction and stopping the machine, according as a quantity computed as a criterion is less or greater than a specified tolerance (M, p. 131).

In addition to this, however, there is the possibility of reading the content of one counter into another, without or with change of sign according to the sign of the number standing in a particular counter (the "choice counter") (M, p. 129); this, however, is a choice between two *numbers*, not between two operations. It could provide a choice between two operations if there were a facility for stepping the sequence control tape on or back between two operations by a number of lines equal to the number standing in a counter. This would give the machine the flexibility of control which Babbage envisaged for the Analytical Engine.

A similar facility was actually provided in a different way, by the addition to the machine of a subsidiary sequence unit (63, p. 29) on which a number of subsequences of computing operations can be set up by plug-in connections. An instruction either on the main sequence-control tape or in one of the subsequences can require that control of the operation of the machine shall be transferred to a specified one of the subsequences, or back to the main sequence control tape. This is a most important modification to this machine, and greatly increases its flexibility.

7.3. Relay Machines

In the Harvard Mark I Calculator, relays are used extensively for controlling the operation of various parts of the machine. In some other machines, relays are used not only for this purpose but for the storage and arithmetical units themselves. Several small special-purpose machines of this kind, and a large general-purpose machine, have been built by the Bell Telephone Laboratories (120, 121). Another large general-purpose relay machine (Mark II Calculator) has been built at Harvard by Professor Aiken (20a).

The first of the B.T.L. machines was completed and demonstrated before the entry of the United States into the war. It is restricted to the operations of arithmetic on pairs of complex numbers; if further calculations are required to be done on the result of one such operation, this result has to be fed back to the machine as one of the data for another operation. Input and output for this machine are expressed in standard teletype code, with a coded symbol for the operation required, so that it can be connected into the teleprinter network of the United States, and can automatically accept data from and supply results to anywhere in the country.

Another machine, known as "relay interpolator," was designed to carry out subtabulation to twentieths, using a cubic interpolation formula (21); it was later found to be adaptable to various other calculations involving the formation of linear combinations of sets of data.

In these machines the storage relays are registers only; all arithmetical operations are carried out in a centralised computing unit. The machines work in a scale of ten, with number expressed in the so-called "bi-quinary" system of coding decimal digits. In this, each decimal digit is represented by a configuration of seven relays consisting of two groups, one of two relays representing 0×5 and 1×5, and the other five representing the digits 0, 1, 2, 3, 4. Any digit is represented uniquely by a configuration in which one, and only one, relay from each group is "on"; this can be used as the basis of a check on the operation of the machine.

Addition is carried out not by counting but by means of an addition table built into the connections between the relays of the computing unit. Control is carried out through instructions coded on the teletype tape.

Fig. 47. General front view of Harvard Mark II Calculator.

The Harvard Mark II (20a) machine was built for the Naval Proving Group at Dahlgren, Virginia. It also is a decimal machine. Control is carried out through punched teletype tapes, in which several rows of holes are read at a time by the reading mechanism. There are three different sequence control mechanisms with facilities for changing over from one to another in the course of calculation, according to a criterion evaluated by the machine. This gives the machine considerably more flexibility of control than that of the original Mark I machine. A front view of the machine is shown in fig. 47.

7.4. The Eniac

The Eniac (39, 50, 102a) was planned by Dr. J. W. Mauchly and Dr. J. P. Eckert and developed at the Moore School of Electrical Engineering of the University of Pennsylvania for the Ballistic Research Laboratory at Aberdeen Proving Ground, where it has now been re-erected.

It operates by counting electrical pulses, produced at the rate of 100,000 per second, by electronic counting circuits, the pulses being routed by electric gates. A standard pattern of pulses is repeated every 20 pulse-periods, that is every 0.2 milliseconds; this is a convenient unit of time in terms of which to express the operation of the machine; being the time required for an addition, it is called an "addition time." The machine works in scale of ten, normally to ten-decimal capacity with a sign indication. In some respects it can be regarded as an electrical analogue of the Harvard Mark I Calculator. The general principle of operation of its arithmetical units can be represented, very schematically and in a very simplified form, by fig. 48. This is derived from fig. 44 by simply relabelling the various components:

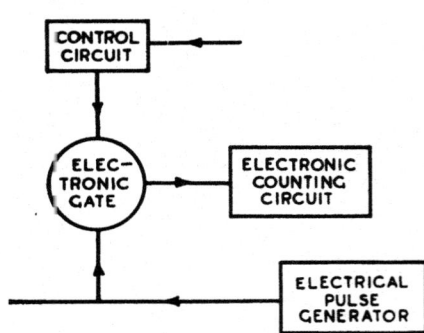

Fig. 48. Basic principle of the Eniac. All connections are electrical.

i. The motor, considered as a source of mechanical impulses, is replaced by an electrical pulse generator.

ii. The mechanical counter is replaced by an electronic counting circuit.

iii. The relay is replaced by an electronic control circuit.

iv. The clutch is replaced by an electronic gate, which only passes pulses from the pulse generator to the counting circuit when the gate is opened by a signal from the control circuit.

The Eniac has twenty electronic counters or "accumulators," each of which serves both as an adding unit and a storage unit, as do the counters in the Harvard Mark I Calculator. Multiplication by *given* small integers can be carried out by repeated addition, but multiplication in general is carried out by use of a built-in multiplication table which forms part of the multiplying unit. A product of two factors, each of one digit, is a number of two digits (either of which may be zero); in a multiplication, the contributions from the left-hand and right-hand digits of the various products of the digits of the two factors are accumulated separately in two accumulators, and then combined. The built-in multiplication table consists of a network such that for any input digits a and b, two groups of pulses from the pulse generator are gated to these two accumulators, one group forming the left-hand and the other the right-hand digit of ab. Multiplication by an n-digit multiplier takes $(n + 4)$ addition times. Division and square root are carried out by a separate unit, the operation of which is comparatively slow.

In the Eniac, the accumulators provide the only form of storage into which numbers can be recorded, and from which they can be read, in times comparable with the times occupied by arithmetical operations. Four of the accumulators, however, are involved in any multiplication, two for holding the factors and two for accumulation of contributions to the product, so that only sixteen are regularly available for storage. The machine has, however, other forms of storage. There are three function tables on each of which 104 values of any function, to 12 places, can be set by hand switches; this process of setting the values is slow, on the time-scale of the machine, but once set, any one is available in a few addition times. Numerical data can also be supplied from punched cards, which are read by a card reader, from which the data are transferred to a set of relay registers; the content of these registers can be changed in the course of the calculation by instructing the machine to feed and read a new card. Results are output on cards through a standard card punch. Both reading and punching a card are slow operations on the time-scale of the machine, requiring 2000 to 3000 addition times, though once a number has been transferred to a relay register, it can be read out in an addition time. Thus although the use of punched cards provides a store of practically unlimited capacity, any extensive use of this for intermediate results slows down the over-all operating speed very considerably, besides requiring the use of an operator to transfer cards from the card punch to the card reader, and perhaps rearrange them in process, so that the whole process of calculation is no longer fully automatic.

The units of the Eniac are interconnected by three sets of lines. One of these is permanently connected to all units, and supplies the standard pattern of pulses produced by the pulse generator. The other connections are set up by hand according to the sequence of computing instructions to be carried out. One set of connections ("digit lines") transfers pulses representing numerical information between the various units; the other set ("program lines") transfers pulses ("program pulses") which stimulate the units to operate in the required sequence. The digits of a number are transferred in parallel on eleven lines (for ten digits and a sign indication), each digit n being represented by a sequence of n pulses. In the connection between a unit and a digit line, an adapter can be inserted giving a shift of any number of places to the right or left, and so dealing with multiplication or division by integral powers of ten.

Figure 49 shows a general view of the Eniac, and fig. 50 shows a closer view of two of the accumulators with connections to the digit and program lines. The tubes and other circuit elements are assembled at the back of the panels.

7.5. Control in the Eniac

Each unit of the Eniac is quiescent until it is stimulated to operate by a pulse input to it from a program line; it then carries out the operation required, and finally a program pulse is emitted by one of the units concerned, and this stimulates the units involved in the next operation. In most operations at least two units are involved, one to transmit and

Fig. 49. General view of the Eniac.

another to receive; when the operation is completed, only one of these need emit a program pulse.

As an example, consider an accumulator. Each of these has five channels by which it can receive from a digit line, and two channels by which it can transmit to a digit line; by one of the latter it transmits the number it is holding, and by the other the complement of this number. It also has twelve channels through which it can be connected to the program lines, and with each of these program channels is associated two switches. One of these switches determines whether the accumulator, when stimulated through this program channel, shall transmit or receive, and by which digit channel, and the other determines whether, after transmission, its content shall be held or cleared. On eight of the program channels there is a third switch which can be set to require the repetition of the process of reception or transmission, up to nine times, before clearing or transmitting a program pulse.

The connections of the various units with the digit and program lines, and the setting of switches such as those on different program channels of an accumulator, form the storage system of the machine for operating

FIG. 50. Two accumulators of the Eniac, and interconnections; the number held in the left-hand accumulator is shown by the indicating lamps above, and is + 42370, 94005.

instructions, and any problem has to be coded in terms of these connections and switch settings.

A unit which is very important for the organisation of a calculation as a whole, but which has not yet been mentioned, is that called the "master programmer." Its purpose is to control the sequence of whole groups of individual arithmetical operations, treating each group as a unit. Most extended calculations consist of such groups of operations, the groups being repeated in a way which it may be possible to specify beforehand but which may depend on the results of the calculation. For example, the step-by-step integration of a differential equation involves the repetition of the sequence of operations for a single interval, which is the main group of operations, but the regular repetition of this group may be interrupted, say every tenth interval, for a record of the current results to be punched; and within this main group of operations there may be an iterative calculation, say for an inverse cube root of one of the variables, which is carried out by repeating a small sub-group of operations inside the main group. In this case, it is known in advance at what stages the regular repetitions of the main group will have to be interrupted, but the number of the repeats of the iterative process may not be known in advance, and may have to be governed by a criterion applied to the results of successive repetitions of the iteration. It is worth noting that successive repeats of the main group may not all involve exactly the same steps, as they may include different numbers of repeats of the iterative sub-group.

The sequence of operations in a single group is determined by the connections between the other units. The sequence of the groups themselves is controlled by the master-programmer, which consists of ten separate six-position electronic switches ("steppers") by each of which an input program pulse can be sent to one or another of six program lines, and so initiate one of six alternative groups of operations, according to the switch position. Each switch can be stepped from one position to the next, or cleared back to its first position, either at predetermined stages in the course of a calculation or in accordance with the result of evaluating some criterion for selection between alternative courses of procedure, such as repeating an iterative process or using the result for the next stage of the calculation. The ten steppers can be interconnected with one another as well as with the rest of the machine, and use of them makes it possible to apply the Eniac to calculations involving a considerable degree of discrimination and selection, the selection being made quite automatically once the machine has been set up.

Since in this form of control the operating instructions have to be supplied to the machine by manual plugging in of the connecting leads and setting of switches, the process of setting up the machine is slow on the time-scale of the machine. On the other hand, it has the advantage that minor modifications of the sequence of operating instructions, if found to be necessary in view of the results of the calculation, can be made and incorporated comparatively easily. In my own small experience of work on the Eniac, I had occasion to appreciate the value of this feature (see §7.9).

7.6. Centralised Control System of the Eniac

The original control system for the Eniac, summarised above, may be described as decentralised, in that elements of it were distributed over the different units of the machine. Recently a form of control from a centralised "control unit," such as Babbage's machine was to have and the Harvard machines have, has been incorporated in the Eniac.

In this new system the plugging of the accumulators into the digit and program lines, and the settings of the switches on the various units, are permanent; further only one accumulator is used as an adding unit, the other, apart from those used in connection with the multiplier, being used as storage units only. The sequence of operations is controlled by an additional unit which is effectively a 100-way switch. It can receive a two-digit number, and give out a program pulse on a corresponding one of 100 output lines. Each of these 100 outputs can be used to initiate a different computing sequence, so that in effect the machine can have 100 different computing routines built into it. The sequence of these routines in any particular calculation is specified by a sequence of two-digit numbers, which are set up by hand on a function table.

This form of control system simplifies the process of setting up the Eniac and checking the set-up, and effectively increases its capacity for operating instructions. It also simplifies testing and trouble-shooting. On the other hand it slows down the over-all operating speed once it has been set up. It seems likely that it will increase the scope and value of the Eniac as a general-purpose machine, but that the older form of control may be better for extensive work on comparatively simple problems such as the step-by-step integration of ordinary differential equations for which the Eniac was originally designed.

7.7. The I.B.M. Selective Sequence Electronic Calculator

Early in 1948 a large general-purpose machine, called the Selective Sequence Electronic Calculator (70), was completed and put into operation by the I.B.M. Company at their headquarters in New York. This has some features of each of the large machines above mentioned, and others of its own.

It has three forms of storage for numbers; a small-capacity, high-speed storage on electronic tubes, a larger-capacity storage on relays, and an indefinitely large capacity on eighty-column paper tape. As in the Eniac, arithmetical operations are carried out through electronic circuits. As in the Harvard Mark II Calculator, the sequence of instructions is punched on a paper tape, and control can be transferred from one to another of a number of sequence control units, one of which can be supplied with the instructions for the main sequence of computing operations and others with instructions for various sub-sequences. The machine is named from the facility it provides for selection between a large number of such sub-sequences. There are altogether 66 locations in the machine at which information on a tape can be read; these can be used for instructions, for tabular material, or for reading numbers previously recorded in the punched-tape storage of the machine for intermediate results.

This machine can be considered as closing the first stage in the development of large automatic calculating machines. The machines considered in this chapter have all been important steps in this development, and, at the time of writing, are still the only ones actually working and producing useful results. However, it seems very improbable that any of them will be duplicated. The machines of the future will be considerably different in principle and appearance; smaller and simpler, with numbers of tubes or relays numbered in thousands rather than the tens of thousands of the machines considered in this chapter, faster, more versatile and easier to code for and to operate. Those at present projected or under construction are different enough to be regarded as forming a second stage of development, and some aspects of them are considered in the following chapter.

7.8. An Application of the Eniac

A short account of an application of the Eniac (24) will serve to illustrate some of its capabilities and limitations.

In the theory of the laminar boundary layer in a compressible fluid, one requires the solutions of a system of ordinary differential equations which can be divided into sets of three. By proper organisation of the work, the different sets of equations can be solved in succession, but the equations of each set have to be solved as three simultaneous equations.

The first set of equations to be solved has the form

$$\left.\begin{aligned}
\frac{df_0}{d\eta} &= \frac{h_0}{[1 + \alpha r_0]^{1-\beta}} \\
\frac{d^2 h_0}{d\eta^2} &= -f_0 \frac{dh_0}{d\eta} \\
\frac{1}{\sigma} \frac{d^2 r_0}{d\eta^2} + \tfrac{1}{2}(\gamma - 1)\left(\frac{dh_0}{d\eta}\right)^2 &= -f_0 \frac{dr_0}{d\eta}
\end{aligned}\right\} \quad (7.1)$$

where α, β, γ, σ are constants; α is the square of the Mach number of the incident stream, β is the index in the relation $\mu \propto \tau^\beta$ between viscosity μ and temperature τ, γ is the ratio of the specified heats, and σ is the Prandtl number. The solution has to satisfy the conditions

$$f_0(0) = h_0(0) = r_0'(0) = 0; \quad h_0(\infty) = 2 \quad r_0(\infty) = 0. \quad (7.2)$$

The general behaviour of h_0 and r_0 is illustrated in fig. 51.

The equations are non-linear, with two-point boundary conditions. There is a convenient iterative method for handling them, but it requires a much greater storage capacity than that of the Eniac (see Ch. 9, §5), so a method of step-by-step integration was used. For such a method, any solution has to be started from definite values of f_0, h_0', h_0, r_0', r_0 at $\eta = 0$; of these three are given but two (h_0' and r_0) have to be estimated and adjusted until the conditions at infinity are satisfied. The situation here is exactly similar to that met in the treatment of differential equations with two-point boundary conditions on the differential analyser (see Ch. 2, §10).

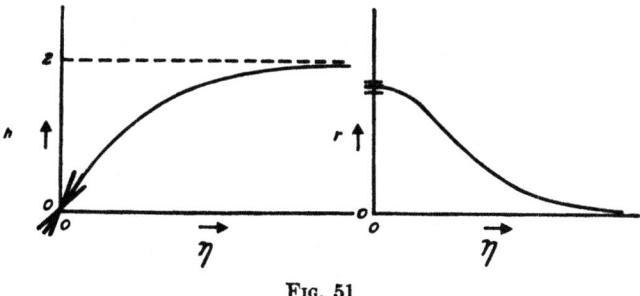

FIG. 51

From the asymptotic form of the solution, it was expected that, to the seven-figure accuracy required, the conditions at infinity would be reached before $n = 5$, so these conditions were replaced by

$$h_0(5) = 2, \qquad r_0(5) = 0. \tag{7.3}$$

From a set of preliminary solutions it was possible to estimate the variations of $h_0(5)$, $r_0(5)$ with the adjustable starting values $h_0'(0)$, $r_0(0)$, and this data could then be supplied to the machine. Then it was possible to set up the Eniac not only to carry out a solution of the equations from given initial conditions but, from the values of $h_0(5)$ and $r_0(5)$ obtained in such a solution, to estimate initial values of h_0' and r_0 for a better solution. Thus the machine alternatively evaluated a trial solution and estimated better initial conditions, until a solution was obtained which satisfies the conditions (7.3) within a specified tolerance. For each of these trial solutions, only one card with initial conditions and final values of $h_0(5)$ and $r_0(5)$ was punched, for record purposes. When a trial solution satisfying the criteria had been found, a final estimation of better initial conditions was made, and a final solution was evaluated, a card now being punched with values of f_0', f_0, h_0', h_0, r_0', r_0 for each interval of the integration.

The whole procedure, including the changes from the operating sequence for the integration to that for estimating better initial conditions, and back, and from punching one card per solution to punching one card per interval, was automatic once the machine was supplied with the data for the solution required. These changes between groups of operations were carried out by the master programmer, which was also concerned in controlling the repetitions of an iterative procedure required to find $1/(1 + \alpha r_0)^{1-\beta}$.

In view of the small storage capacity provided by the accumulators of the Eniac, it was not possible to store many values or differences of the various quantities to be integrated, so a simple integration formula was used, and a large number of small intervals of integration. The interval used was so that 250 intervals were required to cover the range $\eta = 0$ to ε. The machine carried out the integration at the rate of eight intervals a second (not eight a minute, which would have been very fast by any ordinary computing standard, but eight a *second*), so that a single trial solution took about half a minute. The final solution took about $2\frac{1}{2}$ minutes, most of which time was spent in punching cards.

The other sets of equations are of the general form

$$\left.\begin{aligned} f_n' &= \left[h_n - \frac{(1-\beta)\,\alpha h_0}{1 + \alpha r_0}\, r_n \right]\bigg/ (1 + \alpha r_0)^{1-\beta} - C_n \\ h_n'' &= -f_0 h_n' + 2n f_0' h_n - (2n+1)\, h_0' f_n - D_n \\ \frac{1}{\sigma} r_n'' + \tfrac{1}{2}(\gamma - 1)\, h_0' h_n' &= -f_0 r_n' + 2n f_0' r_n - (2n+1)\, h_0' r_n - E_n \end{aligned}\right\} \quad (7.4)$$

where C_n, D_n, and E_n are rather complicated functions of the solutions of equations (7.1) and the solutions f_k, h_k, r_k of the sets of equations of this type with $1 \leqslant k < n$. If these sets of equations are solved in the order on n increasing, C_n, D_n, and E_n are known functions of n when the solution of equations (7.4) is undertaken.

The calculation of these functions can be carried out very effectively by combined use of the Eniac and a reproducing punch. Each solution (f_k, h_k, r_k and their first derivatives) is recorded on a deck of cards, one card for each value of the independent variable. By means of the reproducing punch, those data required for the calculation of C_n, D_n, and E_n can be copied on other decks of cards, which can then be input to the Eniac for the details of the numerical work to be carried out. For example, for D_3 the following quantity is required (though it is not the only contribution):

$$3f_1 h_2' + 5f_2 h_1' - 2f_2' h_1 - 4f_1' h_2. \quad (7.5)$$

f_1', f_2, h_1', h_1 can be copied onto a new deck from the deck giving the functions of order $n = 1$, and f_2', f_2, h_2', h_2 can be copied onto the same deck from the deck giving the functions of order 2, so that for each value of η the eight quantities required for the evaluation of the quantity (7.5) are all on a single card of the new deck. This can then be fed into the Eniac, set up to evaluate (7.5) from such a card input, and the results output on the cards of another deck, which is just the form in which they are wanted as an input in the solution of equations (7.4).

Thus although the attention of an operator is required in this work, this operator is not required to do any calculations with individual numbers, but is only concerned in transferring numbers in large blocks, represented by decks of punched cards, between the different machines. In this way, very powerful use can be made of the Eniac in cooperation with the punched card equipment, particularly the reproducing punch and sorter; and though the process is not purely automatic, it may still be fast, and is certainly labour-saving, compared with other methods of carrying out the calculations.

This work illustrates both the power of the Eniac and its main limitations as a general purpose calculator. The latter are, first, the small capacity of the storage into which numbers can be recorded and from which they can be read in times of the order of an addition time, and secondly the process of making the interconnections by manually plugging in the units to the digit and program lines and setting the switches, which is slow on the time-scale of the machine both to do and to check.

As already mentioned, an iterative method very suitable for equations (7.1) with boundary conditions (7.2) had to be discarded on account of the small storage capacity available, and for the same reason the integration had to be carried out using simple integration formulae and a large number of intervals. The process of making the interconnections will be simplified by the new centralised control system outlined in §7.6, but the sequence of operating instructions must still be set up by hand on the program table and checked, and this is still a slow operation on the time-scale of the machine.

It should be added that the Eniac was originally designed primarily for the step-by-step integration of the differential equations of external ballistics, though its control was made flexible enough for it to be used for other calculations within its capacity. For this primary purpose its capacity is adequate and its method of setting up is not inappropriate, for in such work the main interconnections will remain the same over a long period of time and the quantities needing adjustment from one set of runs to another, such as switch settings for values of the ballistic coefficient, the interval of integration, etc., are very easily set. In this context the limitations mentioned are hardly noticeable; it is only when the machine is considered as a general-purpose machine that they become at all marked.

7.9. The "Machine's-Eye View" in Programming a Calculation

It has been mentioned that in the work outlined in §7.7 an iterative method was used for the evaluation of $1/(1+\alpha r_0)^{1-\beta}$. In connection with this there arose a point which, although trivial in itself, seems of some importance as an example of a general principle.

For any value of $\alpha r_0 = z$, the first approximation to $1/(1 + z)^{1-\beta}$ to use in the iterative calculation was taken from a function table, and was such that two repeats of the iteration formula were certainly enough to give $1/(1 + z)^{1-\beta}$-to the accuracy required in the subsequent work. Since r_0 was to be always positive, this table was only set up to cover positive values of z; but in doing this the point was overlooked that although *in the solution finally required* r_0 is always positive, it might happen that in one or more of the *trial* solutions, r_0 would become negative. And this did in fact occur.

Now the argument to this function table is taken as the first two digits after the sign indication in the accumulator holding z, so that when r_0 went negative and z occurred in this accumulator as $\bar{1}.99\ldots$ ($\bar{1}$ representing the sign digit), the approximate value of $1/(1 + z)^{1-\beta}$ was read out for $z = 0.99$ instead of for $z = -0.01$; and then two repeats of the iteration were by no means sufficient to give $1/(1 + z)^{1-\beta}$ to the accuracy required, so that further results were spurious.

A human computer, faced with this unforeseen situation, would have exercised intelligence, almost automatically and unconsciously, and made the small extrapolation of the operating instructions required to deal with it. The machine without operating instructions for dealing with negative values of z, could not make this extrapolation. To put it descriptively, the machine didn't know what to do with a negative value of z; it hadn't been told. It did its best, and this was something quite sensible according to its structure and the limited instructions which *had* been given to it, but quite different from the correct small extrapolation of these instructions which a human computer would make as a matter of course.

It must not be concluded that the machine *could* not be given instructions to deal with the situation. If this had been foreseen, then either the table of $1/(1+z)^{1-\beta}$ could have been extended to small negative values of z, or instructions could have been programmed to tell the machine how to deal with negative values of z if they occurred. The latter course was in fact taken when the situation was realised as a consequence of the anamolous behaviour of some of the results; and the ease with which the necessary modifications of the set-up were made has already been mentioned at the end of §7.5. The point is that the machine could *not* deal with the situation *without* specific instructions for the purpose; as this situation had not been foreseen, these instructions were not supplied; and consequently spurious results were obtained.

The moral of this experience is that in programming a problem for the machine, it is necessary to try to take a "machine's-eye view" of the operating instructions, that is to look at them from the point of view of the machine which can only follow them literally, without introducing anything not expressed explicitly by them, and try to foresee all the unexpected things that might occur in the course of the calculation, and to provide the machine with the means of identifying each one and with appropriate operating instructions in each case. And this is not so easy as it sounds; it is quite difficult to put oneself in the position of doing without *any* of the hints which intelligence and experience would suggest to a human computer in such situations.

Chapter 8

PROJECTS AND PROSPECTS

8.1. The Main Directions of Development

As already mentioned, the two limitations of the Eniac, in its original form, as a general-purpose machine were the small capacity of its high-speed storage and the manual process by which the sequence of operating instructions was set up; the machine carried out individual arithmetical operations at high speed, but had not the facilities to make full use of this speed except in calculations for which its small storage capacity was adequate. These two points indicate the main directions in which the first further developments are to be expected, namely the provision of a form of high-speed storage of much greater capacity without a corresponding increase in electronic equipment, and a means by which the machine can set up for itself the connections required for the sequence of computing operations, as in the Harvard Mark I calculator and various relay machines, instead of these connections having to be set up manually in advance.

These two aspects are related, since an instruction, like a number, can be coded in the form of an ordered set of 0's and 1's, and then the kind of high-speed storage used for numbers can also be used for instructions. And if this were done, adequate capacity for instructions as well as for numbers would have to be provided in a single storage system.

Instructions may in any case consist partly of numbers, specifying locations in the store, so that a coded digital form for instructions is a convenient one to use. Then there need be no difference in form between the representation in the machine of a number and an instruction; the difference is in the way they are used. And even in this there may not be a sharp distinction, since if the instructions are stored in this form it is possible to carry out arithmetical operations on them, such as adding a number to the part of an instruction specifying a location in the store. This facility may sometimes be a very valuable one in economising storage space for instructions, and for simplifying programming; hence the use of a numerical notation for the storage locations, as well as being convenient, is not a trivial matter.

8.2. Storage Systems

The development of a simple reliable form of high-speed storage, with large capacity, is of major importance for the further development of large automatic machines. At least three forms of storage have reached a state of development at which they can be used for such a machine. These are: (i) acoustic delay lines, in which data are stored in the form of trains of acoustic pulses, (ii) magnetic material in the form of wire, tape, or the

94

surface of a drum, on which data are stored in the form of variations in the magnetic state of the material, and (iii) insulating screens on which they are stored in the form of a static charge distribution.

A simple acoustic delay line (99, 100, 113) consists of a tube of mercury with a quartz piezo-electric crystal at each end. An electrical pulse, input at one end, is converted into a supersonic acoustic pulse (perhaps on a carrier) in the mercury, which is used in preference to any other fluid as it provides the best acoustic match to quartz. At the other end of the tube the acoustic pulse is converted again into an electrical pulse, which is then amplified and could be fed back to the transmitting end so as to circulate through the delay line and associated circuits. But a pulse group

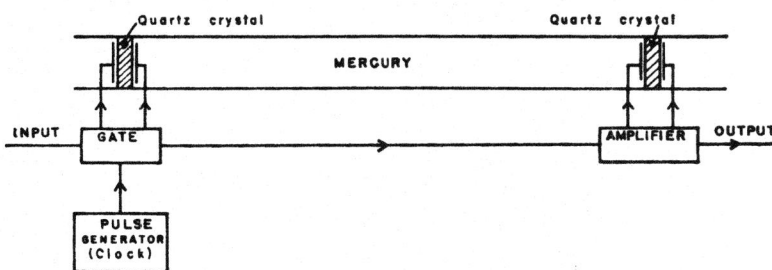

FIG. 52. Supersonic acoustic delay line.

may be required to circulate many thousands, or even millions, of times before being used, and progressive degeneration of the pulse shape arising from distortion between the input to the delay line and amplified output from it would lead to the pulse becoming unrecognisable. So the amplified pulse is not itself transmitted, but used to open a gate so as to allow a freshly-made clean-shaped pulse from a pulse generator to reach the transmitting end of the delay line (see fig. 52). At a pulse spacing of 1 microsecond, 150 cms. of mercury will hold about 1,000 pulses, which is enough to represent, in binary form, 30 numbers of 10 decimal digits each, and the information in such a length becomes available once each millisecond. The number of electronic tubes required to give this storage capacity is quite small, so that it is possible to obtain a considerable storage capacity without an amount of equipment which would be prohibitive from the points of view both of capital cost and of maintenance.

If a number of such delay lines are used, the time-differences between the delays provided by the separate lines must be kept to a small fraction of the interval between pulses; this requires accurate location of the quartz crystals in construction, and a close approach to temperature uniformity in use. If one line is used to control the frequency of the pulse generator, the actual value of the temperature can be allowed to float, provided uniformity of temperature is secured; but if the pulse generator is of fixed frequency, the temperature must be controlled to a fixed value. One way of simplifying the problem of attaining uniformity of temperature is to use a number of delay channels in a single container of mercury.

The magnetic material used in magnetic storage may be in the form of wire, tape, or the surface of a cylindrical drum, which is usually called, for brevity, a "magnetic drum" (72). In any case, the material passes under a "writing head" which produces a magnetic field and magnetises the material in accordance with the data to be stored, and a "reading head" which produces an electrical signal depending on the magnetic state of the material being passed under it. In some cases the same head may be used for writing and for reading, and a number of heads may be used to write or read at different positions on a single wire or channel.

A wire can carry only one channel of information; a small number of channels can be used on a tape and a large number on a drum. Thus different means have to be used for identification of a storage location on a wire and on a drum, and this affects the means of using these different forms of magnetic storage. Also the question of access to stored information is different for the different forms of storage; information on a drum running at constant speed automatically becomes accessible at regular intervals of time, whereas a wire or tape has to be scanned for the required information.

Various forms of electrostatic storage are under development and preliminary accounts of two of these have been published (34, 93); these both involve the use of special electronic tubes. Another form, which has the advantage of using standard cathode ray tubes, has been developed more recently by Professor F. C. Williams of the University of Manchester, England (118). In this, the screen of the tube is scanned in a series of parallel lines, and each line is divided into a series of intervals on each of which one of two charge distributions can be placed. The screen is backed by a metal plate which picks up a signal of a sign depending on which kind of charge distribution is being scanned, and this signal can be used to read out the information stored in the form of the charge distribution, and also to control the beam intensity so as to regenerate the charge distribution. Writing is carried out by an alternative control on the beam intensity.

There is no need to restrict a single machine to a single form of storage, and several machines being developed at present make use of two forms, a "high-speed" storage of fairly small capacity—perhaps 200 numbers— which has direct connections with the arithmetical unit, and into which numbers can be recorded and from which they can be read in times of the order of the times required for arithmetical operations, and a "slow" storage of much greater capacity, with facilities for transfer between the two forms of storage (these two forms of storage are sometimes referred to as "internal" and "external," or as "main" and "auxiliary," respectively).

8.3. Serial Machines Using Delay-Line Storage

One group of machines at present under development uses delay-line storage both for numbers and instructions. It includes, in the United States, the EDVAC (Electronic Discrete Variable Automatic Calculator) at the Moore School of Electrical Engineering of the University of Pennsylvania, the Univac and Binac being built by the Eckert-Mauchly Computer Corporation, Philadelphia, and in England the A.C.E. (Automatic Computing Engine) at the National Physical Laboratory (115) and the EDSAC (Electronic Delay Storage Automatic Calculator) at the Mathematical Laboratory of the University of Cambridge (114).

Although, as pointed out in §5.7, use of a dynamic form of storage does not impose the use of a serial representation of numbers or a serial method for arithmetical operations, it does suggest a serial treatment, and all these machines are in fact serial in character. Also all use two-indication signals, of which the two indications can be interpreted as digits 0 and 1. But in various other respects they differ considerably from one another. The Univac is a decimal machine working in the "added-three" code; the others all normally work in scale of two, though the A.C.E. is intended as a universal machine and will be able to be programmed to work in scale of ten — or any other scale — and this may also be the case for the others. The Univac also has a large auxiliary storage in the form of magnetic tape. They also use different standard forms of instruction, with corresponding differences in the organisation of the control system and in methods of programming, and they also differ in the extent and form of built-in automatic checking equipment.

In such machines, it is convenient to group the series of digital positions into groups of standard length, which are called "words" whether they refer to numbers or instructions. Typical lengths of words are 32, 36, 40 or 48 digital positions. The term "minor cycle" is used for the period of time occupied by a "word," and "major cycle" for the delay time of a storage delay line. A "double-length" number or instruction is one which is two words long; for example in multiplication, the full product of two numbers each of one word length is a double-length number.

8.4. Functional Analysis of Serial Machines

The analysis of the operation of a machine using two-indication elements and signals can conveniently be expressed in terms of a diagrammatic notation introduced, in this context, by von Neumann and extended by Turing. This was adapted from a notation used by Pitts and Mc-

Culloch (80) as a possible way of analysing the operation of the nervous system, which is also composed of "two-indication" elements, in that an individual neuron is only capable of an "all-or-none" response, not of a graduated response.

An element has an output line on which it can give out a signal, and one or more input lines on which it can be stimulated, the stimuli either exciting it to emit a signal or inhibiting it from doing so. The character, exciting or inhibiting, of the stimulus produced by a signal on any one input line is fixed. An element regarded, ideally, as emitting a signal instantaneously on excitation, is represented by a circle with a number in it (see fig. 53), the number indicating the least number of simultaneous excitatory stimuli required to stimulate the emission of a signal. An element for which this number is n may be called a "threshold-n" element. An inhibitory stimulus overrides any excitation; an input line on which a stimulus has an inhibiting effect on an element is indicated as shown in fig. 53 (d).

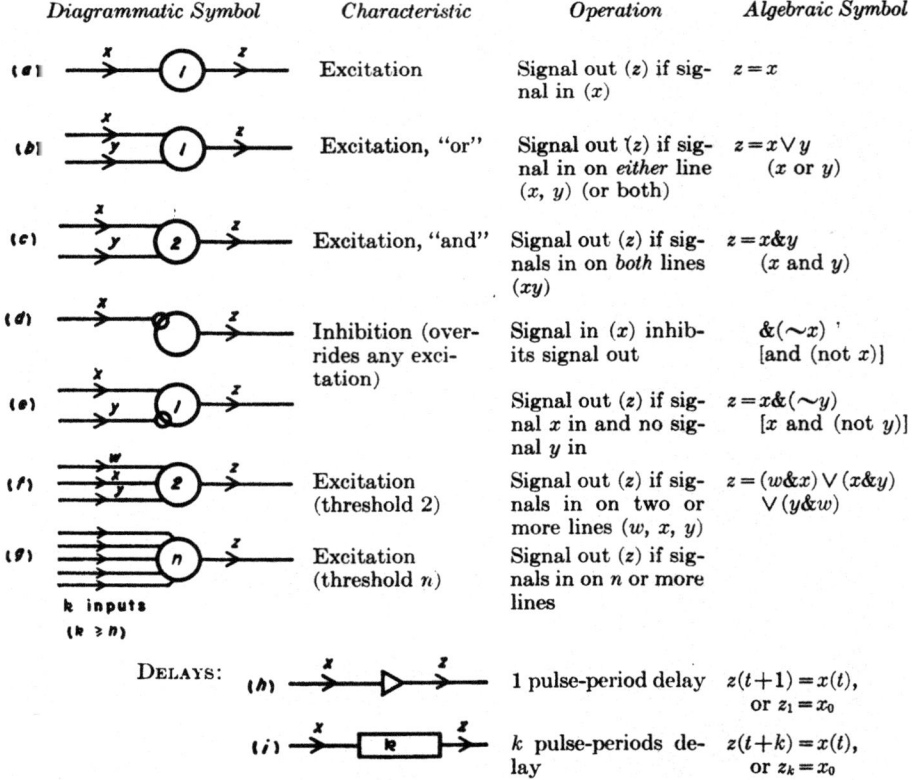

Diagrammatic Symbol	Characteristic	Operation	Algebraic Symbol
(a)	Excitation	Signal out (z) if signal in (x)	$z = x$
(b)	Excitation, "or"	Signal out (z) if signal in on *either* line (x, y) (or both)	$z = x \vee y$ (x or y)
(c)	Excitation, "and"	Signal out (z) if signals in on *both* lines (xy)	$z = x \& y$ (x and y)
(d)	Inhibition (overrides any excitation)	Signal in (x) inhibits signal out	$\& (\sim x)$ [and (not x)]
(e)		Signal out (z) if signal x in and no signal y in	$z = x \& (\sim y)$ [x and (not y)]
(f)	Excitation (threshold 2)	Signal out (z) if signals in on two or more lines (w, x, y)	$z = (w \& x) \vee (x \& y) \vee (y \& w)$
(g)	Excitation (threshold n)	Signal out (z) if signals in on n or more lines	

k inputs
$(k \geq n)$

DELAYS: (h) — 1 pulse-period delay — $z(t+1) = x(t)$, or $z_1 = x_0$

(i) — k pulse-periods delay — $z(t+k) = x(t)$, or $z_k = x_0$

FIG. 53. Functional elements of a serial machine.

The important kinds of elements for the present context are those represented in fig. 53 (b, c, e, h, i). These are not all independent; those represented in fig. 53 (b, c) can be built up from elements of the character represented by fig. 53 (e). And those represented by fig. 53 (f, g) can be built up from the simpler elements, fig. 53 (b, c). In the present context, threshold-3 elements are seldom involved, and threshold-n $(n > 3)$ elements very seldom if ever.

In a serial machine, "words" are indicated by sequence of signals "0" or "1," occurring at equal intervals; as these signals are often pulses, this interval will be called the "pulse period." Essential elements in such machines are (i) an element for giving a delay of one pulse period, and (ii) one giving a delay of k pulse periods. Unless k is small, a delay of k pulse periods will probably be obtained by some other means than a series of delays each of one pulse period, so it is convenient to regard these two kinds of delay elements as different. They are represented as shown in fig. 53 (h, i).

The elements represented in fig. 53 (a-g) are idealised in that the response, on application of the appropriate stimulus, is supposed to be immediate. This is a departure from the usage of Pitts and McCulloch, who regarded each element as introducing a delay; this was necessary in their neurological context. In the present context, however, it is a useful idealisation to regard excitation and delay as being functions of separate elements.

The analysis of a machine into elements of this kind is a functional rather than a structural analysis. The circuit used to obtain the properties of a single element may consist of several tubes or other circuit components, or on the other hand the functions of two or three such elements may be combined in a single tube or simple circuit. The designer and the maintenance engineer will need diagrams more closely representative of the actual physical hardware. But for the user who is more concerned with what the machine does than the details of the circuity and mechanism by which it does it, analysis into these functional elements seems to me more illuminating and easier to follow than detailed or block circuit diagrams.

There is a form of algebra which is very suitable for expressing relations between a set of connected elements each of which is capable of only two indications. This is Boolean algebra, which has already been used by Shannon (98) for the analysis of relay and switching circuits. In the present context, a symbol (such as x) stands for the indication provided by an element or signal, the possible "values" for this symbol being 0 and 1, 1 being represented by the excitation of an element or the presence of a signal on a line, and 0 by the contrary indication. In this algebra, the essential operations are those expressed by the words "not" (applied to a single quantity), "or" and "and" (applied to a pair of quantities).*

---

* "Or" here is the so-called "inclusive or"; "x or y" (written $x \vee y$) is 1 if either x or y or $both$ are 1.

(a)	x	0	1		
	$\sim x$ (not x)	1	0		

	x	0	0	1	1
	y	0	1	0	1
(b)	$x \vee y$ (x or y)	0	1	1	1
(c)	$x \,\&\, y$ (x and y) xy	0	0	0	1
(d)	$x \wedge y$ $[x \cdot (\sim y) \vee (\sim x) \cdot y]$	0	1	1	0
(e)	$x + y$ (x plus y) (single r.h. (sum) digit is ($x \wedge y$) digits) l.h. (carry) digit is ($x \,\&\, y$)	00	01	01	10

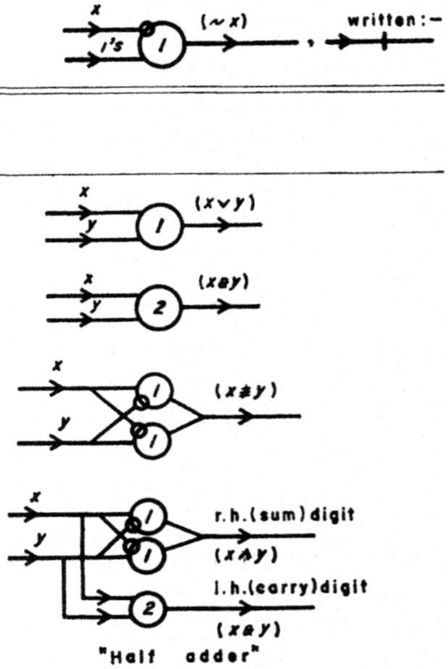

Fig. 54

Another operation, which can be constructed from these, is that which gives the result 0 when x and y are the same, and 1 when x and y are different, that is to say, if either x or y *but not both* are 1;[†] this will be written $x \wedge y$. In terms of the operations "not" ($\sim$), "or" ($\vee$), and "and" ($\&$),[‡] it is defined by

$$x \wedge y = [x \,\&\, (\sim y)] \vee [(\sim x) \,\&\, y]. \tag{8.1}$$

It is easily verified that this operation is commutative (that is, $x \wedge y = y \wedge x$) and associative [that is, $x \wedge (y \wedge z) = (x \wedge y) \wedge z$] so that it is unambiguous to write $x \wedge y \wedge z$ without brackets, and further that

$$\left.\begin{array}{l} x \wedge y \wedge z = 1 \text{ if one or three of } x, y, z \text{ are } 1 \\ \qquad\qquad = 0 \text{ otherwise.} \end{array}\right\} \tag{8.2}$$

This relation is important in connection with binary addition.

The significance of these operations, and the realisation of them in terms of the elements of fig. 53, is given in fig. 54; and in the right-hand column of fig. 53, the operations of these elements are expressed in terms of this algebra.

---

† This is the so-called "exclusive or."

‡ Since the operation "and" has many of the properties of multiplication, the symbol & is often omitted, and "x and y" written simply as xy.

It will be noted that the elementary operations of Boolean algebra (sometimes called "logical operations") are simpler than the operations of arithmetic. Arithmetical addition of two numbers of one digit each may lead to a number of *two* digits, whereas the operations of Boolean algebra all give results of one digit only. If binary addition of single digits x and y is considered as *always* giving a number of two digits, either one of which may be zero, the two digits may be expressed as follows in terms of these "logical operations":

$$\left.\begin{array}{l}\text{[left-hand (carry) digit of } x+y] = x \,\&\, y \\ \text{[right-hand (sum) digit of } x+y] = x \wedge y\end{array}\right\} \tag{8.3}$$

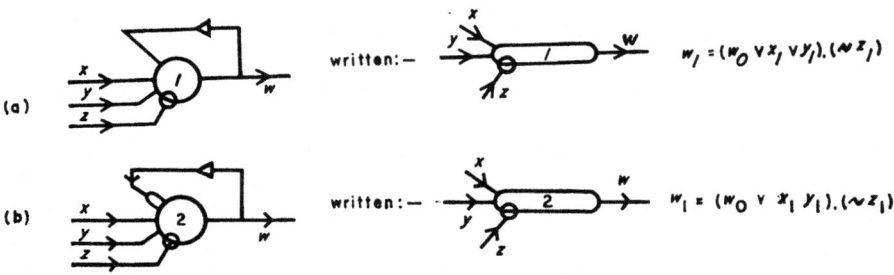

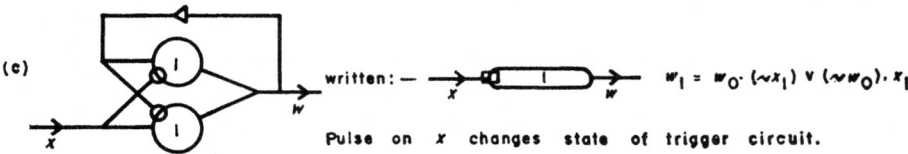

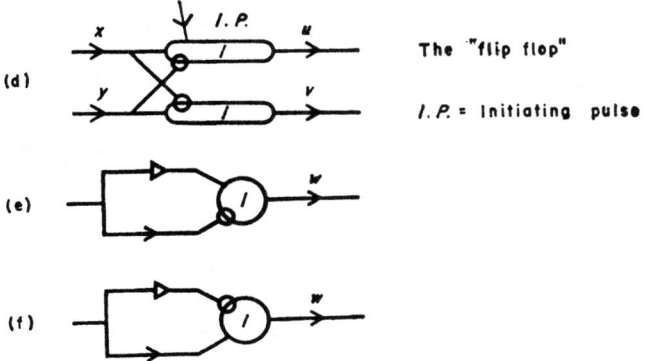

Fig. 55. (a-d), trigger circuits; (e), circuit for indicating end of a train of 1's; (f), circuit for indicating beginning of a train of 1's.

(see fig. 54, e). An element which constructs these two outputs from inputs x and y is known as a "half-adder," since, as will be seen shortly, an adder for numbers expressed in serial binary form can be constructed from two half-adders with suitable interconnections.

Other elements can be built up from those represented in fig. 53, and are used often enough to make it convenient to introduce single symbols for them; these extensions of the notation are due to Turing. Some of these elements are illustrated in fig. 55. The most important of these are the trigger circuits. The connections to these may be of three kinds, exciting, inhibiting, or reversing; a reversing connection is one on which a stimulus changes the state of the trigger circuit, whether it was previously "off" or "on," it is indicated as shown in fig. 55 (d). Figure 55 (e, f) shows elements one of which signals the beginning and the other the end of a sequence of input pulses.

Another most important element is a storage element, which can be thought of as a delay line with terminal equipment for transmitting and clearing the data contained in it, and for accepting new data. This, and the abbreviated symbol for it, is shown in fig. 56. If there is no signal on the control line D, no signal at x can pass the threshold-2 element A into the delay line, but there is no inhibitory stimulus on the threshold-1 element B, so the output from the delay line is fed back to the input end of the line. Thus under these conditions the content of the delay line circulates through the delay line and the element B. If during a certain period there is a signal on the control line D, this inhibits the feed-back of the output of the delay line to its input end, and therefore clears from

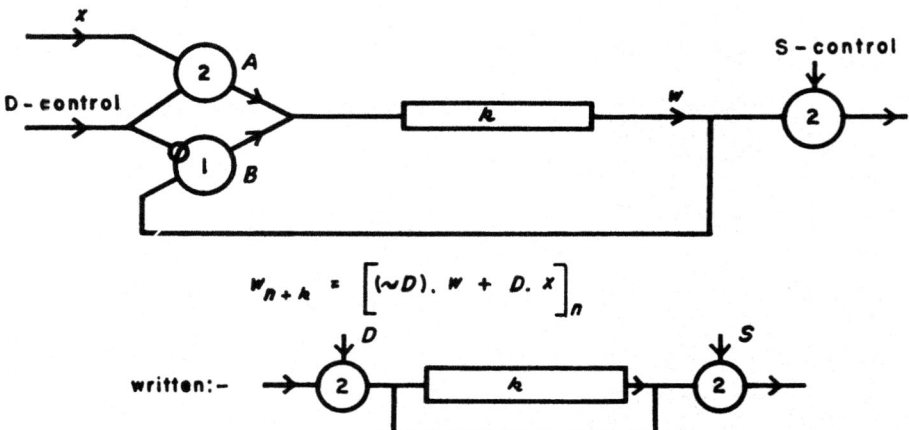

$$w_{n+h} = \left[(\sim D) . w + D . x \right]_n$$

FIG. 56. Delay element with terminal equipment to control reception, transmission, storage, and clearance (storage element).

the storage unit the information which is output during this period; at the same time it opens the "gate" formed by the threshold-2 element A, so that any signals at x are accepted by the storage unit. Normally, information is only cleared in the process of accepting other information in its place. Information is read out from the storage element by a signal on the control line S, which opens the "gate" formed by the threshold-2 element C.

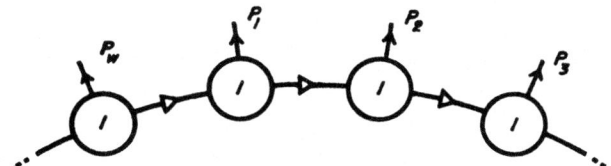

FIG. 57. Pulse-counting ring (clock).

A necessary component of a serial machine is a clock, both to count pulse periods within a minor cycle so that any digital position in a word can be identified if required, and to count minor cycles within a major cycle. A unit which combines the functions of a pulse generator and pulse counter is a ring of threshold-1 elements connected by delays of 1 pulse period each (see fig. 57). The signals emitted by the elements will be called $P1$, $P2 \ldots Pw$ where w is the length of a word. Gating of clock pulses P can be used as a means of maintaining synchronism between the pulses in corresponding digital positions in different parts of the machine. A simplified form of adding unit can be used as a minor-cycle counter, and will be considered in the course of §8.5.

8.5. Arithmetical Operation

In the addition of two binary numbers we have to cater for the carry-over from each place to the next most significant. If the numbers are expressed in serial form with the *least* significant digit leading, the carry-over can be obtained by means of a delay of one pulse period. If x_n, y_n are the digits of the addends x and y in the n-th place from the right, c_n the carry digit into this place, and s_n the digit in this place in the sum $x+y$, the rules of binary addition are expressed by:

$$s_n = (x \wedge y \wedge c)_n \qquad (8.4)$$

$$c_{n+1} = (xy \vee yc \vee cx)_n. \qquad (8.5)$$

There are several ways of building up an adding unit, for numbers expressed in this way, out of elements of the kind considered in §8.4. One, using two half-adders, is shown in fig. 58 (a). The first half-adder forms the sum-digit and carry-digit from the digits x_n and y_n *alone*, and the second deals with the carry-over process. The analysis of the operation of this adder in terms of the algebraic representation of the operation of the various elements is interesting as an example of the use of Boolean algebra in this context.

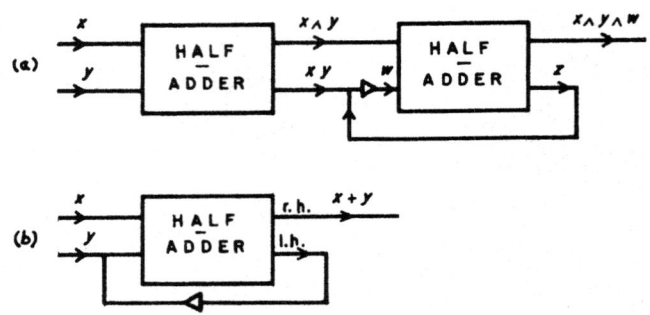

FIG. 58. (a) Synthesis of adder from two half-adders. (b) Simplified adder when one addend (y) has only one non-zero digit.

For digital position n, let z_n be the output in the carry-digit of the second half-adder. The outputs from the first half-adder are $(x \wedge y)_n$ and $(x\bar{y})_n$ [see fig. 54 (e) and formulae (8.3)]. The inputs to the second half-adder are $(x \wedge y)_n$ and

$$w_n = [(xy) \vee z]_{n-1}, \tag{8.6}$$

and its outputs are $(x \wedge y \wedge w)_n$ for the sum-digit and

$$z_n = [w(x \wedge y)]_n \tag{8.7}$$

for the carry-digit. Then it follows from (8.7) that $(xyz)_n = 0$, so that there are never simultaneous pulses on the two lines into the delay element. Also elimination of z gives

$$w_n = [xy \vee w(x \wedge y)]_{n-1};$$

this simplifies to

$$w_n = [xy \vee yw \vee wx]_{n-1},$$

which is the recurrence relation satisfied by the carry-over digit c_n in the sum $(x+y)$ [see formula (8.5)]. Now for the least significant figure, say $n = 0$, $z_0 = 0$ since there is at most one input to the second half-adder. Hence $w_1 = (xy)_0 = c_1$, the carry-digit into the next to least significant figure. Hence $w_n = c_n$, and the output $(x \wedge y \wedge w)_n$ of the second half-adder is s_n [see formula (8.4)].

If the output from such an adder is connected back to one of its inputs (say y) through an element giving a delay of 1 minor cycle (fig. 59), it will accumulate the sum of successive numbers input at x, and is called a "cumulative adder."

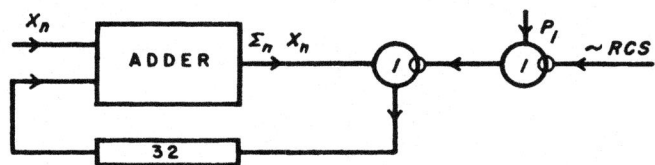

FIG. 59. Cumulative adder with control of round-carry. (Round-carry is normally suppressed, but if $\sim RCS = 1$, this suppression is inhibited.)

If one of the addends (say y) is 1 in a single binary place only, then addition can be carried out by a single half-adder (fig. 58b), as there can never be simultaneously a pulse on y and a carry pulse. A cumulative adder simplified in this way, with the clock pulse Pn as its input, adds 1 in the n-th place for every minor cycle, so that it forms a minor cycle counter, counting in terms of the n-th digit as unit (fig. 60).

Another form of adder, due to von Neumann, is represented in fig. 61 (a). A cumulative adder can be formed from it by feeding the output back as one of the addends, as in fig. 59, and if x is 1 in a single binary place only, the threshold-3 element is never required, so the adder can be simplified by omitting it. A minor cycle counter consisting of a cumulative adder simplified in this way is shown in fig. 61 (b).

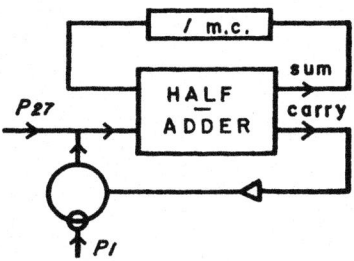

FIG. 60. Minor cycle counter (simplified cumulative adder) counting on $P27$.

A multiplier is a special form of cumulative adder, in which the input is controlled to be zero or the multiplicand, according as the digit of the multiplier is 0 or 1, and the delay line is two minor cycles and one pulse period long, the double length to hold the full number of digits of the product, and the one extra pulse period to give the shift between the contributions from successive digits of the multiplier.

Since in a serial machine the adder is a comparatively simple unit, it is practicable to incorporate two adders and do each addition in duplicate. The results can then be compared and the next operation only undertaken if this check is satisfactory.

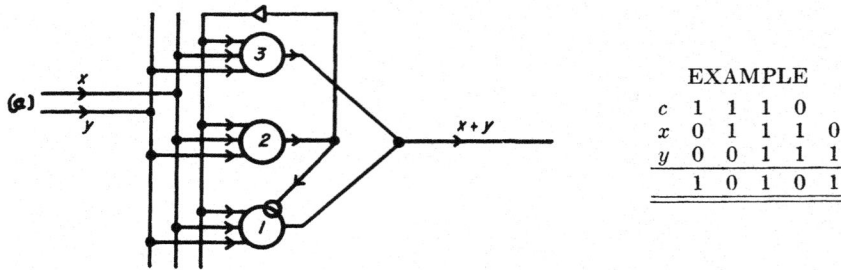

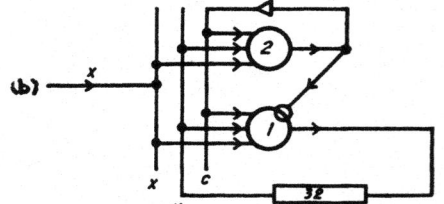

EXAMPLE

c	1	1	1	0	
x	0	1	1	1	0
y	0	0	1	1	1
	1	0	1	0	1

Minor cycle counter (modified adder)

$x = 0$ except for one digit of each minor cycle.

$(x = P1, \text{ for example})$

FIG. 61. Alternative form of binary adder.

8.6. Control

The function of a control system is to take the operating instructions in the appropriate order and to take the action necessary to ensure that each is carried out. How it does this will depend considerably on such characteristics of the operating instructions as whether they are of "one-address" or "four-address" form (see §5.8) or of some other form, and what criteria can be used for conditional selection of the next instruction. Different machines differ considerably from one another in these respects, and it does not seem possible in a short account to cover them all. One possible organisation of a control system will be considered here as an example. It is not the control system of any actual machine, though it illustrates some features of the control system of the A.C.E.

Most operations in a delay-line machine involve the transfer of a word from or to a delay line, and the control system must be able to select the delay line to be used as source or destination for each transfer, and to time the transfer so that it starts at the beginning of the right minor cycle and continues for the right period.

In the illustrative example here considered, it is supposed that a standard operation is a transfer from a specified source S to a specified destination D, and that the standard instruction specifies the source and destination numbers, and a "timing number" t, and that the period for which the transfer continues is specified by a "characteristic" c in such a way that when $c = 0$ the transfer starts in the minor cycle after the instruction has been received by the control system, and continues for t minor cycles, whereas if $c = 1$ nothing happens for t minor cycles, and then the transfer starts and lasts for a single minor cycle. The instructions are located not in serial order in a delay line, but in such a way that when the operation specified by one of them has been carried out, the next is the one which at that instant is just about to become available as input to the control system. Thus it is only necessary to specify explicitly the delay line in which this instruction is located; the selection of the appropriate word in the delay line is ensured by the timing number and the ordering of the instructions in the store; planning the latter is part of the process of coding a problem for the machine (see §8.9).

A possible organisation of the transfer circuits for such a machine is illustrated in fig. 62. Here a number of source-gates formed by threshold-2 elements on the right, and a number of destination-gates on the left, are connected by a single bus labelled "Highway." In this bus there is a further gate, labelled "transfer gate," which exercises the main control over transfer of words between the various sources and destinations. An instruction specifies one source and one destination, and, when received

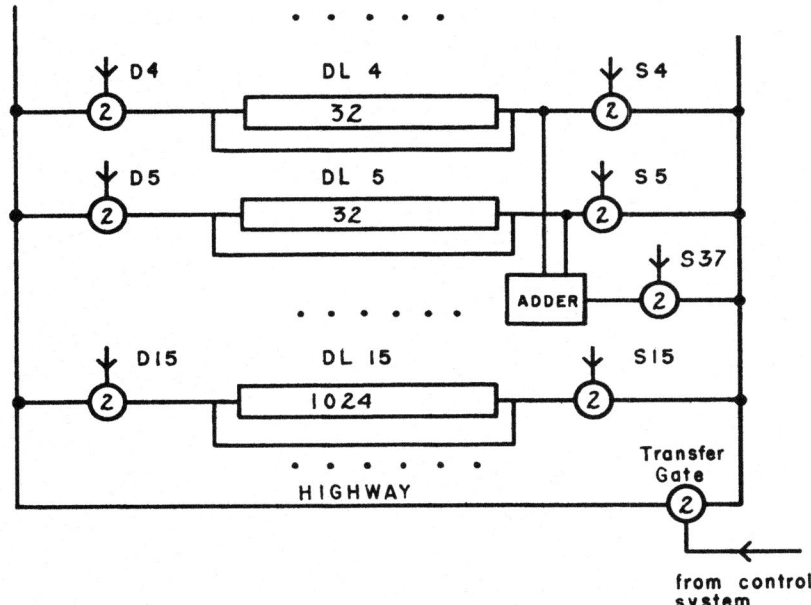

FIG. 62. Illustrative example of transfer circuits.

by control system, opens the corresponding *S*-gate and *D*-gate through selecting circuits; then at the appropriate time the main transfer gate in the highway opens and remains open for the number of minor cycles required for the transfer.

An organisation for the main control system to carry out these operations is shown in fig. 63. This consists of five main units, three of them, enclosed in broken lines in fig. 63, being assemblies of functional elements of the kinds shown in figs. 53 and 54, and two trigger circuits labelled *TC* 1 and *TC* 2. Of the three larger units, one is a static register or "staticiser" on which the instruction currently being carried out is set up in static form, one controls the conditional selection of the next instruction, and the third is a minor cycle counter (see figs. 60b, 61b). The trigger circuit *TC* 1 controls the transfer of numbers; when it is "on," a signal from it opens the main transfer gate in the highway (see fig. 62) and allows the transfer, specified by the source- and destination-numbers set up on the staticiser, to proceed. The trigger circuit *TC* 2 controls the reception of the instructions by the control system.

This system operates as follows. Suppose first that *TC* 3 is off. Let us start from a stage in which *TC* 1 is "on" so that a transfer is taking

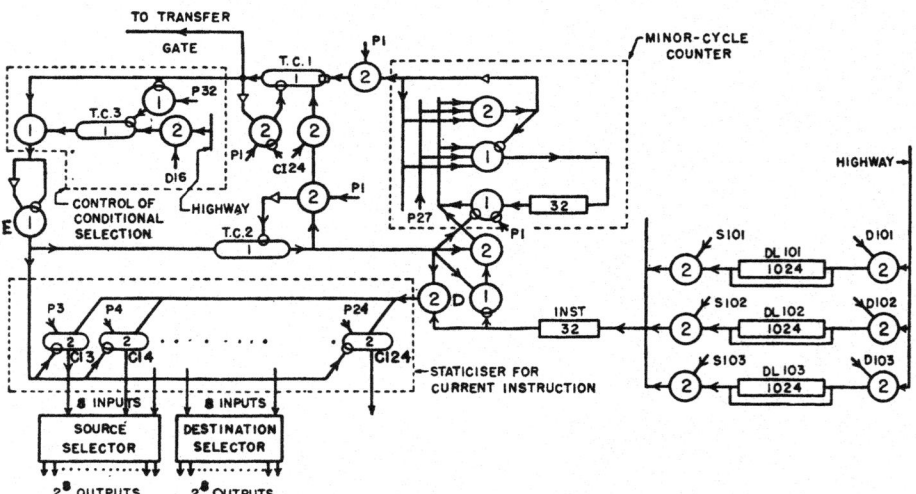

FORM OF INSTRUCTION.—SOURCE NO. ; DESTINATION NO.; CHARACTERISTIC ; TIMING NO.

DIGITS 3–12   13–22   24   27–32

TRIGGER CIRCUIT {
T.C.1 CONTROLS TIMING OF TRANSFER BETWEEN SOURCE AND DESTINATION
T.C.2 CONTROLS TIMING OF SETTING UP NEXT INSTRUCTION
T.C.3 CONTROLS CONDITIONAL SELECTION OF NEXT INSTRUCTION
}

FIG. 63. Illustrative example of functional organisation of control circuits.

place; it will appear shortly that $TC\,2$ is necessarily "off" at this stage. When the transfer required is complete, $TC\,1$ is put off, and this can only occur at the first pulse ($P1$) of a minor cycle. The end of the signal output by $TC\,1$ results in a single pulse being emitted from the element E, which both clears the staticiser and excites $TC\,2$, which remains on for just one minor cycle, being extinguished by a pulse on its inhibitory input at $P1$ of the next minor cycle. While it is on, the gate formed by the threshold-2 element D is open, and digits 3-24 of the word w then being emitted from delay line "INST" will be set up on the staticiser; the instructions must be so located that this word is the next instruction to be acted on. Signals from $TC\,2$ also clear the minor-cycle counter and allow the complement of w to be set up in it. The first 26 digits of this are irrelevant; the counting is done in digits 27-32.

Digit 24 of the instruction is the "characteristic" already mentioned; if this is 0, trigger circuit $CI\,24$ of the staticiser is not excited, and $TC\,1$ is excited by the pulse emitted by $TC\,2$ at the end of the minor cycle when it is on; the transfer starts immediately, and continues until $TC\,1$ is extinguished by a pulse on its reversing input; this occurs when there is a carry-over from the most significant digit of the content of the minor-cycle counter. If, on the other hand, the "characteristic" is 1, trigger circuit $CI\,24$ is excited, and its output inhibits the excitation of $TC\,1$ by the last pulse output by $TC\,2$. Then $TC\,1$ remains off until it is excited by a pulse on its reversing input, and it is then extinguished after one minor cycle.

Now suppose that $TC\,3$ is on. Then the train of pulses input into E lasts for one minor cycle after $TC\,1$ is extinguished, so the next instruction is received by the staticiser one minor cycle later than it would be if $TC\,3$ were off. Thus conditional selection of the next instruction is obtained by sending to destination $D\,16$ (the input to $TC\,3$) a 0 or 1 according as the relevant criterion is or is not satisfied, and locating the two alternatives for the next instruction next to one another in a delay line.

As already mentioned, this must not be regarded as a representation of the control system of an actual machine, but as a simplified illustration of some of the operations a control system must be able to do and one kind of way in which it can be organised to do them.

8.7. Parallel Machines

A type of machine far removed from the delay-line machines considered in the previous sections is one using a static form of storage and parallel operation. At present the only practical form of large-capacity static storage seems to be some form of electrostatic storage of charge distribution.

The main differences of such a machine from a delay-line machine are:

(i) No precise timing of operations is needed, either in the operation of the machine or in the programming. The content of any storage location is available at any time, and the completion of the operation specified by one instruction is all that is needed to initiate the reading of the next instruction;

(ii) In the arithmetical unit, a separate component must be provided for addition in each binary place (or in each digital position in a coded decimal machine), and carry-over must be obtained by some means other than a time delay;

(iii) Shifting (multiplication by 2 in a binary machine) must also be accomplished by other means than a time delay.

As with the series machines, there are various possible ways of carrying out the various functions which have to be provided in such a machine. A possible form of adding unit, built up of a set of half-adders connected together, is shown in fig. 64; if this is drawn out in more detail, it will be seen that the connections through the threshold-2 units (see fig. 54 (e)) of the right hand set of half-adders carry out an operation equivalent to the "simultaneous carry-over" of Babbage's Analytical Engine and some other machines (see §5.4).

Multiplication can be performed by successive additions of the multiplicand with appropriate shifts, each individual addition of the multiplicand being a parallel operation; the question of carry-over is simplified if multiplication is carried out in the order of *increasing* significance of

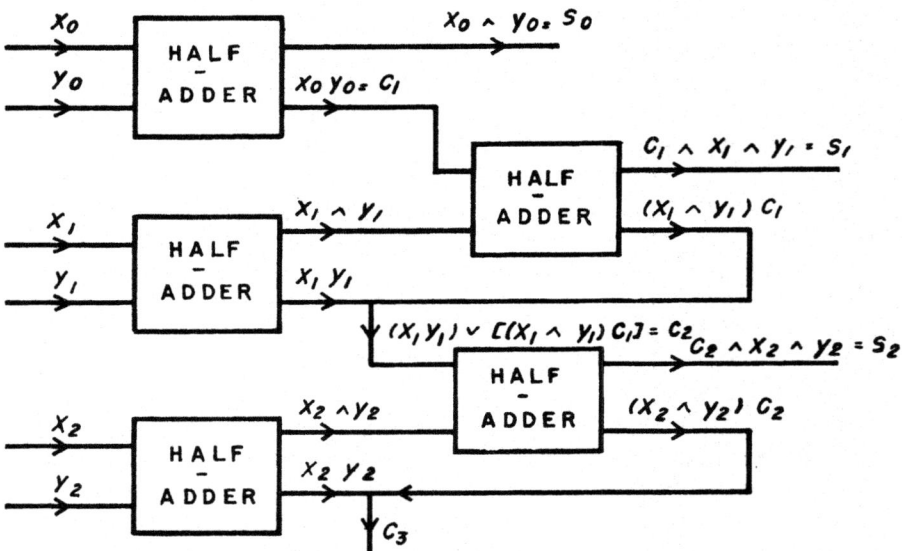

FIG. 64. Synthesis of parallel adder from half-adders.

the digits of the multiplier; then after p digits of the multiplier have been used, the last p of the digits of the product will not be affected in subsequent stages of the multiplication, and can be stored in a part of the machine which is purely a register, without carry-over facilities; thus a full double-length product can be formed without requiring that the adding-unit proper be able to handle more than single-length numbers.

The control system of a parallel machine with static storage is likely to be considerably simpler than that of delay-line machine, possibly to such an extent as to offset the additional complication in the arithmetical unit due to the need to duplicate components in order to obtain parallel operation.

A machine of this kind, with electrostatic storage supplemented by magnetic wire, is under construction at the Institute of Advanced Study, Princeton.

8.8. Other Types of Machine

Another type of machine is one which uses magnetic material as the main storage, rather than as an auxiliary storage as in the Univac and in the Princeton machine. If the magnetic material is in the form of a magnetic drum, information can be stored on a number of parallel channels round the drum, and synchronism between these channels is assured by the rigidity of the drum, and does not depend on precise control of speed. One or more of the channels can be used to carry index signals, and the position of the drum at any moment can be identified by keeping a count of these index signals as they pass under a reading head.

Such a drum can be used either for serial or parallel storage of "words." In a series form of storage, the digits of a word would be recorded serially on a single channel; in a parallel form, they would be recorded in similar positions, relative to the reading heads, on several channels. It is also possible to use a mixed form, for example in a coded decimal machine the four or five digits representing each decimal digit could be coded in parallel on four or five channels, but the successive decimal digits could be stored serially. How it would be used depends on the organisation of the rest of the machine with which it is associated. If the arithmetical unit is intended for parallel operation, it would probably be most convenient to use a parallel form of storage on the drum.

Two machines using magnetic storage are under construction, the A.R.C. (Automatic Relay Computer) by Dr. A. D. Booth in England (12a), which, as its name implies, will have a parallel relay arithmetical unit, and "Mark III Calculator" by Professor H. H. Aiken at Harvard.

8.9. Programming and Coding

The process of preparing a calculation for a machine can be broken down into two parts, "programming" and "coding." "Programming" is the process of drawing up the schedule of the sequence of individual operations required to carry out the calculation, and "coding" is the process of translating these operations into instructions in the particular form in which they are read by the machine.

Newcomers to the subject often seem to find difficulty in appreciating the degree to which a calculation has to be broken down into individual operations before it can be coded for a machine. A machine can only carry out the operations of arithmetic, transfer operations, possibly some of the "logical operations" (see §8.3), and discrimination; any calculation must be broken down into a sequence of such operations before it can be handled by the machine. I have been asked, for example, if the machine can construct a table of prime numbers in a specified range, say 10^7 to 2.10^7. The answer is "Yes, if you can specify a sequence of these operations which will lead to the required result." The machine has no inbuilt knowledge of prime numbers and their properties, or direct means of recognising them; it can only obtain them and identify them by certain procedures which have to be programmed. Lady Lovelace's comment (see p. 70) "The machine can only do what we know how to order it to perform" is as appropriate now as it was a hundred years ago.

Although the various kinds of machines considered in the previous sections differ considerably from one another in their internal organisation and operation, the general process of programming a calculation will be much the same for any of them, for it depends primarily on the structure of the sequence of operating instructions required to carry out the calculation. Some characteristic features on an individual machine will, however, affect the details of the programming. Such features are:

(i) The standard form of operating instruction adopted; whether this is, for example, a "one-address" or "four-address" form (see §5.8).

(ii) The facilities provided by the standard instructions; for example, whether division can be carried out directly or has to be done by means of an iterative process which has to be programmed.

(iii) The criteria which it is possible to use for discrimination between positive alternative courses of procedure; for example, whether it is only possible to discriminate on the sign of a number or also on other criteria such as the likeness or unlikeness of signs of two numbers.

Von Neumann and Goldstine (40) have proposed a method of indicating the structure of the sequence of operating instructions by means of a "flow diagram" representing the control sequence. This is in the form of a block diagram, in which the blocks represent operations or groups of operations, and are joined by directed lines representing the sequence of these operations.

A simple example is provided by the iterative process

$$x_{n+1} = x_n[10 - ax_n^2]/9$$

for $a^{-1/9}$ (see §9.2), x_n^9 to be obtained as $(x_n^3)^3$ by repeated cubing, and the quantity $z = (x_{n+1} - x_n)^2 - \epsilon$ to be evaluated as a criterion of whether the iterative process had been carried far enough, that is, whether x_{n+1} can be used as an adequate approximation to $a^{-1/9}$ or whether a further repetition of the iterative process is required. The flow diagram for this process is shown in fig. 65.

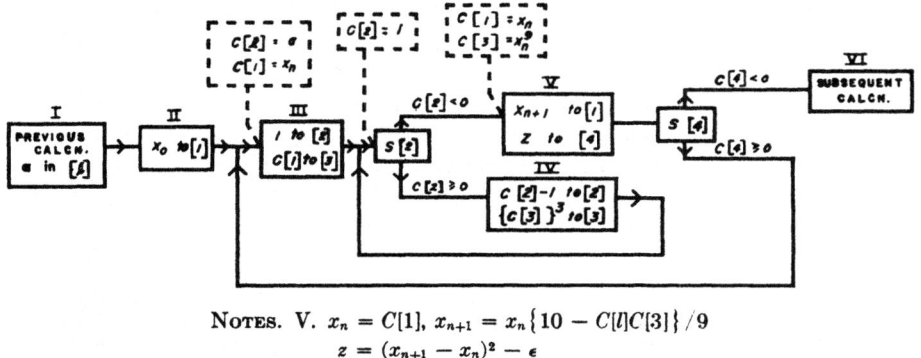

NOTES. V. $x_n = C[1]$, $x_{n+1} = x_n \{10 - C[l]C[3]\} / 9$
$$z = (x_{n+1} - x_n)^2 - \epsilon$$

FIG. 65. Block diagram of process of finding $a^{-1/9}$ by iteration formula
$$x_{n+1} = x_n [10 - ax_n^9]/9.$$

A block labelled with a roman numeral, and with one input and one output, represents an arithmetical, logical, or transfer operation. Formulae representing the operation, or groups of operations, may be written in the corresponding blocks, or listed separately; the former is more convenient in the actual process of programming, though the latter may be a more compact way of presenting the results. A block with two outputs represents a selection between two alternative procedures; the symbol in such a block stands for the quantity which is being used as a criterion to discriminate between these alternatives. It is convenient to represent a number l specifying a storage location by $[l]$, and the content of this location by $C[l]$*; and the sign of the content of l by $S[l]$.

The blocks outlined by broken lines in fig. 65 do not represent operations of the control, but contain explanatory notes and memoranda for the programmer; in the actual process of programming they can conveniently be distinguished by the use of red ink or pencil.

---

* This is an addition to von Neumann and Goldstine's notation which I have found convenient. In programming more involved sequences of operations, it may be convenient to write $C^2[l]$ for the content of the storage location whose number is $C[l]$, the number which is at storage location l, and $L[x]$ or $C^{-1}[x]$ for the number of the storage location whose content is the value of a quantity x in the formula to be evaluated.

In the example shown in fig. 65, the loop through the operation block IV is concerned with the evaluation of y^3 from y, and $C[2] = i$ is a number used to count the repetitions of this calculation. At the beginning of the $(n+1)$-th repetition of the iterative process, $C[1] = x_n$ and $C[2] = 1$; $C[2]$ is decreased by unity for each repetition of the cubing process, so that after two repetitions of this process, when $C[4] = x_n^9$, $C[2]$ becomes negative and the criterion $S[2]$ then results in the control selecting the first operation in block V as the next to be carried out. The other selection, between repeating the iteration and going on to the subsequent calculation, is dealt with similarly.

To make the machine as flexible as possible, it seems advisable to have considerable freedom in the selection of the criterion to use for discrimination between two alternative courses of procedure. It is possible to use $S[l]$ as the only criterion, but it is convenient also to be able to discriminate on the basis of other criteria, such as the equality of two numbers, or perhaps the parity of a number or the likeness of the signs of two numbers without reference to their magnitude.

Another facility which is likely to be very useful in operation is that of stopping at any selected point in a calculation and then inspecting the content of any storage location. Both the EDVAC and Univac will have this facility, the "breakpoint" in which the sequence of operations is stopped being set by hand switches so as to be under full current control by the operator. They also have facilities for manual setting of a word on a set of switches, so that after inspection of results at a breakpoint, the operator can decide whether to continue the normal course of the calculation or perhaps to go back to an earlier stage or to skip some instructions and go on to a later stage. These facilities are likely to be particularly valuable in exploratory work, when the user cannot see far ahead how the calculation is going, but wants to inspect some results and decide on the basis of them what to do next. They are likely also to be valuable in testing and checking both the machine and the programming.

Chapter 9

HIGH-SPEED AUTOMATIC DIGITAL MACHINES AND NUMERICAL ANALYSIS

9.1. Introduction

Our present standard methods of carrying out extended numerical calculations are based on the capabilities of equipment at present generally available, or which has been available in the past. With the development of new kinds of equipment of greater capacity, and particularly of greater speed, it is almost certain that new methods will have to be developed in order to make the fullest use of the capabilities of this new equipment. It is necessary not only to design machines for the mathematics, but also to develop a new mathematics for the machines. We must try to look ahead and foresee some of the possible methods and problems which will have to be studied in connection with the application of the new machines which we may expect to become available in the next few years.

The following are some of the properties we may expect of one of these machines. It may be expected to carry out individual arithmetical operations in times of the order of a millisecond for each operation; to have a capacity of the order of some thousands of "words," each of which may be either a number or an instruction, and to take the instructions in succession and set up automatically the proper connections between its component units. Also if at any point in the calculation there are two (or more) alternative courses for the subsequent calculation, it will select and carry out the correct one according to specified conditions.

The speed makes it practicable to use methods of calculation involving large numbers of single arithmetical operations, whereas our present methods are aimed at saving such operations. For example, if multiplications can be carried out at the rate of a million per hour, which is already possible with the Eniac, a calculation involving 10^5 multiplications is almost trivial as far as the time taken to carry out the numerical work is concerned; the time taken to program the calculation would probably be much larger. And calculations involving 10^7 or even 10^8 multiplications are by no means too large to be considered.

In general the time taken to carry out the detailed numerical work with such a machine, relative to the time taken to plan it, is likely to be very much less than it would be without the machine. Hence whereas without the machine it may often be worth spending some time developing methods for cutting down the amount of numerical work involved in a calculation, with a machine it may be more effective to use simple crude methods rather than spend time developing more elaborate methods involving fewer arithmetical operations.

An example occurred in the application of the Eniac to the equations considered in §7.8 (see ref. 24). These were differential equations with two-point boundary conditions, which were treated by evaluating a number of trial solutions. Two methods were available using the results of previous trial solutions to estimate better initial conditions. One was crude but simple to apply, and needed very little programming and only a small amount of quite rough calculation which could be quickly done beforehand; the other was more accurate, and would enable the required solution to be reached with fewer trials, but was more elaborate and would have taken much longer to program. The cruder method was preferred, since each trial solution took only half a minute, and the extra time taken by the greater number of trial solutions was less than the extra time it would have taken to program the more elaborate process. Thus the cruder method was actually the more effective in using the machine, whereas without the machine, it would certainly have been worth while using the more elaborate method in order to reduce to a minimum the number of trial solutions to be evaluated. This example shows that though experience with hand computation (assisted by desk machines) is valuable in planning calculations with high-speed automatic machines, it must be used with discretion and with appreciation of the great difference between the conditions of calculation without and with the high-speed machines.

The large number of arithmetical operations which may occur in the course of a single calculation raises rather acutely the question of accumulation of rounding-off errors — the errors resulting from the use of a finite number of significant figures at each stage of the calculation. The study of these errors and their effects is one of the general investigations required on account of the capabilities of these new machines.

9.2. Iterative Methods

A class of methods which seems very suitable for mechanisation is that of iterative calculations (30, 55).

If an equation $f(y) = 0$ is written

$$y = F(y) \tag{9.1}$$

a solution may be found by constructing the sequence $\{y_n\}$ defined by

$$y_{n+1} = F(y_n); \tag{9.2}$$

if this sequence tends to a limit Y, then $y = Y$ is a solution of the equation. The process of forming the successive quantities y_n is called *iteration*, and these quantities are called successive *iterates* to $y = Y$; such a method is called an iterative method. Such a method can also be used in cases in which the unknown y is an ordered set of numbers, a matrix, or a function of a continuous variable x, or perhaps of several such variables. When y is a function of x, the equation corresponding to (9.1) is $y = Oy$ where O is an operator which may involve differentiation or integration. Such methods are suitable for mechanisation for two reasons; they are to a certain extent self-checking, and the successive repetitions of the iterative process make little demand on storage capacity.

Iterative methods can be classed according to the way in which the error $\eta_n = y_n - Y$ of the n-th iterate decreases as the number n of repetitions of the iterative process increases. If y is a *number*, as distinct from a set of numbers or a function, expansion of the right-hand side of equation (2) in a Taylor series about $y = Y$, gives

$$\eta_{n+1} = a_1 \eta_n + a_2 \eta_n^2 + a_3 \eta_n^3 + \ldots \tag{9.3}$$

where

$$a_k = F^{(k)}(Y)/k!$$

An iterative method in which $a_k = 0$ for $k < j$, $a_j \neq 0$ may be called a "j-th order" iterative process. For a first-order iterative process, the number of iterations required to obtain each new correct significant figure in the result remains the same; for a second-order process the number of correct significant figures is doubled for each iteration; for numerical work, a second-order process is *very* much better than a first-order process.

When y is a number it is always possible to derive a second-order process from a first-order process as follows (3, 4, 55, 97). If y_0, y_1, y_2 are three successive iterates of a first-order process, let a quantity Y^* be derived from them by the formula

$$Y^* = \left[y_0 y_2 - (y_1)^2 \right] / (y_2 - 2y_1 + y_0). \tag{9.4}$$

Let Y_{n+1}^* be the quantity obtained by putting $y_0 = Y_n^*$ in this formula. Then the process of forming the successive values of Y_n^* is second-order. For practical work it is not convenient to calculate Y^* from this formula, since it expresses Y^* as the ratio of two quantities each of order η_0, which may be small. But if it is written so that Y^* is expressed as the sum of the best available approximation y_2 and a correction

$$Y^* = y_2 - (y_2 - y_1)^2 / (y_2 - 2y_1 + y_0), \tag{9.5}$$

the correction itself is of order η_0, and there should be no difficulty in evaluating it. The extension of this to cases in which y is a set of numbers or a function of one or more independent variables would be an important step. When y is a single number it is usually fairly easy to set up a second-order iterative process directly, but when y is a set of numbers or a function, it is usually not obvious how to set up anything better than a first-order process, and some method of improving the results of such a process is badly needed.

Consideration of formula (9.3) gives the convergence of the iterative process when the error η is small, but another requirement is a method of improving the convergence in the early stages when the error may be still quite large. It is possible that this would have to be worked out afresh for each particular problem.

A simple example of an iterative process is one for finding a reciprocal without doing any division; this is useful because division (in scale of ten, at any rate) is an untidy process to mechanise. If a is given, the sequence y_n defined by

$$y_{n+1} = y_n (2 - ay_n) \tag{9.6}$$

converges to $1/a$ (provided y_0 lies between 0 and $2/a$); use of this formula gives a second-order iterative process. A third-order process can be based on the formula

$$y_{n+1} = y_n [3 (1 - ay_n) + (ay_n)^2]. \tag{9.7}$$

Formula (9.6) is a special case of the formula (see ref. 62, p. 176)

$$y_{n+1} = y_n [(p + 1) - (ay_n^p)]/p \tag{9.8}$$

which gives a second-order iterative process for $a^{-1/p}$, and a similar generalisation of (9.7) to give a third-order process for an inverse p-th root can be obtained. Another important special case is given by $p = 2$, which gives $a^{-\frac{1}{2}}$, from which $a^{\frac{1}{2}}$ can be obtained by multiplying by a.

Alternatively, a useful direct second-order process for a square-root $a^{\frac{1}{2}}$ is given by

$$y_{n+1} = y_n (3a - y_n^2)/2a. \tag{9.9}$$

This may be more convenient in practice than the better-known second-order process based on Newton's formula

$$y_{n+1} = \tfrac{1}{2} (y_n + a/y_n) \tag{9.10}$$

since use of (9.9) involves only one division, to give $1/a$, whereas use of (9.10) requires a new division for each repetition of the iteration.

Iterative methods can sometimes be used effectively for the evaluation of solutions of differential equations; examples are considered later (§§9.5, 9.6).

9.3. Simultaneous Algebraic Equations

It is probably the case that there is no one best method for the evaluation of the solution of a set of simultaneous linear algebraic equations, but that the best method in any particular case depends on the structure of the set of equations concerned. This is certainly true of methods for treating such equations without the use of an automatic machine, and may well still be true when such assistance is available. For example, it may be that most of the coefficients are non-zero, and are not small integers and moreover are known only approximately, either because they are derived from measurements which are subject to experimental error or because they are results of previous calculations and are subject to rounding-off errors; or it may be that each equation involves only a few of the variables, and these with coefficients which are small integers and are known to be exact. It is quite likely the most appropriate methods in the two cases will be different.

The standard textbook method is the use of the expressions for the solution as the ratios of determinants, but the use of this form by direct evaluation of the determinants is quite unsuitable as a practical computing method. Practical methods can be divided into two classes, and have been distinguished by the terms "direct" and "indirect" (38). Indirect methods are those in which the unknowns are obtained by a process of successive approximation, as in the "relaxation method" of Southwell (35, 36, 101, 102). In such methods the computing process used is applied repeatedly, and the number of repetitions required to obtain a specified approximation to a solution is not predetermined. Direct methods are those, such as the use of elimination, in which the unknowns are obtained in a single application of the process, to any accuracy, provided that an adequate number of figures have been retained in the work.

Some form of successive elimination may be the most convenient direct method with an automatic machine. Such methods have come under suspicion lately on account of an estimate by Hotelling (68) of the possible cumulative effects of rounding-off errors. However, a discussion and comparison of various direct methods have recently been carried out at the Mathematics Division, N.P.L. (38); these have shown that although it may be possible deliberately to construct systems of equations to which this estimate applies, experience with many sets of equations not so constructed indicates that for general application it is much too pessimistic. An analytical study of rounding-off errors in the elimination process has also been made by von Neumann and Goldstine (88).

By a slight extension, the method of elimination can be used to invert the matrix of the coefficients, and if solutions for a set of equations

$$\Sigma_j a_{ij} x_j = b_j \qquad (9.11)$$

with given coefficients a_{ij} are wanted for a number of sets of values of b_i, the most efficient method may be first to evaluate an approximation, say a^*, to the inverse of the matrix a of the coefficients, and then to multiply each set of b's by a^*. Since, on account of rounding-off errors in the calculation of a^*, this matrix will not be exactly the inverse of the matrix a, the result ξ_j will not be exactly a solution of the equations (9.11), though it should be a good approximation to a solution; it is advisable therefore to calculate the quantities $c_i = (b_i - \Sigma_j a_{ij} \xi_j)$ and solve in a similar way the equations

$$\Sigma_j a_{ij} (x_j - \xi_j) = c_i$$

(compare 4.1a) for the corrections to the approximate solution ξ_j.

Of the indirect methods, Southwell's "relaxation" method is very convenient for pencil-and-paper work with equations of suitable form. But although it should be possible to devise a special-purpose digital machine for carrying out the process, it does not seem suitable for a general-purpose automatic machine of any of the kinds at present contemplated, as the unit process consists of scanning a large number of quantities, for which these machines are not particularly suited and would take a comparatively long time, and then carrying out a very simple calculation selected on the basis of the results of this scanning. Moreover, an experienced worker can hasten the process of approximation by trying to anticipate at each step the effects of later steps in the process. It might be possible to study the way in which the human computer exercises judgment in this context and to analyse it far enough to incorporate a simple form of it in the operating instructions to the machine; but this would probably require quite an involved set of discriminations and conditional selections of instructions, which would make substantial demands on storage.

Another process depends on the fact that, for a set of x_j which *is* a solution of the equations, each of the quantities

$$r_i = (\Sigma_j a_{ij} x_j - b_i)$$

which Southwell calls the "residual" for the i-th equation, is zero, so that $\Sigma_i r_i^2 = 0$, whereas for any set x_j which is *not* a solution, not all the r_i's are zero, so that $\Sigma_i r_i^2$ is positive. So the determination of a solution is equivalent to minimising the quantity

$$S = \tfrac{1}{2} \Sigma_i r_i^2,$$

and we can try to obtain the solution by minimising S by a direct numerical process.

At any point x_j, it is possible to evaluate the direction of the greatest rate of decrease of S, that is, the direction of $-(\text{grad } S)$. This is normal to the surface $S = \text{const.}$ through x_j (see fig. 66), and is given by

$$dx_j = -k\frac{\partial S}{\partial x_j} = -k\Sigma_i r_i\frac{\partial r_i}{\partial x_j}, \qquad (9.12)$$

where k is independent of j and is positive. This forms a set of equations for a "curve of steepest descent," that is a curve of which the tangent at each point is in the direction of $-(\text{grad } S)$ at that point. Evaluation of a solution of these equations from any starting point x_j should lead to a minimum of S and so to a solution of equations (9.11); such a method has been called a "method of steepest descent."* Continuous integration

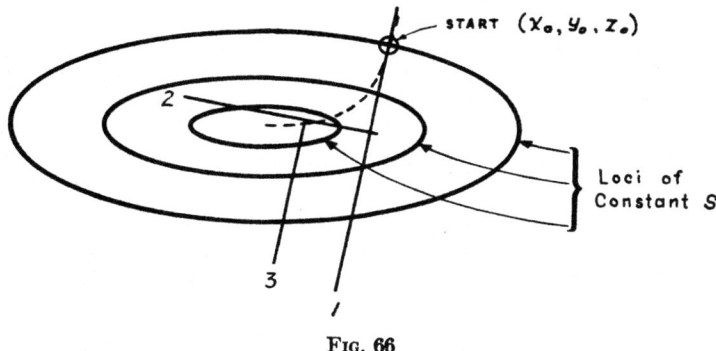

Fig. 66

of these differential equations has been used as a means of solving sets of simultaneous algebraic equations on the differential analyser (103).

A step-by-step solution of the differential equations for a curve of steepest descent would be possible on an automatic digital machine, but such a curve is only a means of reaching the minimum of S and is of no further interest or value, so that no purpose is served in calculating it accurately; and in any case the first step in such a process is the replacement of derivatives by finite differences, so that an alternative process not so closely related to the differential equations may be more effective. One alternative is to evaluate S at a number of points x_j along the direction of steepest descent at some point $(x_j)_{(0)}$ and from these to estimate the position $(x_j)_{(1)}$ of the minimum along that line (fig. 66, line 1); then determine the direction of steepest descent at $(x_j)_{(1)}$ (fig. 66, line 2), and so on. Some experiments I have made with this method (54) suggest that

* The term is used here in a different sense from that in the theory of functions of a complex variable.

it may not be quite as straightforward as it sounds, but Booth (12) has reported a successful use of it in another context.

An important advantage of both these indirect methods is that they can be applied to non-linear equations, and Booth's application of the method of steepest descent was in fact to a set of such equations.

9.4. Solution of Ordinary Differential Equations

It has already been emphasised in §2.10 that in the practical evaluation of the solutions of differential equations, either by the differential analyser or by numerical process, the nature of the boundary and other conditions is much more important than the form of the equation; and, concerning the boundary conditions, what is most important is not *what* they are but *where* they are. There are two common forms, one in which all the boundary conditions are specified at one end of the range of integration, and the other in which some conditions are specified at one end of the range and some at the other end, or in which there is a relation, such as a condition of periodicity, between the conditions at the two ends. Another type of condition which sometimes arises is a condition on the solution as a whole, such as a normalising condition $\int_a^b y^2 dx = 1$, or a condition of orthogonality between two functions in the solution of a set of simultaneous equations, as in the example:

$$y'' + [f_1(x) - \lambda_1]y + [g(x) - \lambda_{12}]z = 0$$
$$z'' + [f_2(x) - \lambda_2]z + [g(x) - \lambda_{12}]y = 0$$

with conditions $y = z = 0$ at $x = a$, $x = b$, and

$$\int_a^b y^2\, dx = \int_a^b z^2\, dx = 1, \qquad \int_a^b yz\, dx = 0.$$

If all conditions are boundary conditions at a single value of x, integration can start from there with nothing unknown, and usually proceeds without difficulty, apart perhaps from the occurrence of singularities. But if some boundary conditions are given at one end of the range of integration and some at the other, or if there is some condition on the solution as a whole, then it is usually necessary to estimate some conditions at one end of the range, or one or more parameters in the equation itself, or both, and to evaluate a set of solutions with different values of these estimated initial conditions and parameters, and to adjust these estimates until a set is found for which the solution satisfies all the specified conditions. In special cases this may not be necessary; for example, for a single linear homogeneous equation a normalising condition can be satisfied by evaluating any solution satisfying the boundary conditions and multiplying by a normalising constant afterwards. And if the equations are linear but inhomogeneous, with no parameter in the equations themselves to be determined, it is formally possible to obtain the solution required by superposition of a particular solution and a complementary function; however, it is very often found that this is only a purely formal possibility, and is quite unsuitable for practical numerical work. And in general these procedures are not available.

9.5. Solution of Ordinary Differential Equations With One-Point Boundary Conditions

For integration of differential equations with one-point boundary conditions, step-by-step methods seem quite satisfactory. In such a method, the range of integration is divided into a set of equal intervals δx, and if the solution has been taken to $x = x_0$, the next step is to carry the integration through an interval δx to $x = x_1 = x_0 + \delta x$.

If the differential equation is of the order n, and all the functions occurring in it can be differentiated as often as required, a powerful method (15) is to differentiate the equation m times and so obtain formulae for the derivatives up to the $(m+n)$-th in terms of those up to the $(n-1)$-th, and then use the Taylor series at $x = x_0$. If S_e and S_0 are the sums of the even and odd terms in this series, then $y_1 = y(x_0 + \delta x) = S_e + S_0$, $y_{-1} = y(x - \delta x) = S_e - S_0$. The latter provides a very good check on the integration from y_{-1} to y_0 and on the formulae used for calculating the derivatives of x_0. Similar formulae can be used for the derivatives up to the $(n-1)$-th.

If some of the functions occurring in the equation are specified by a table of values and not by an analytical formula, such a method is not usually applicable on account of the difficulty of obtaining satisfactory values of the higher derivatives of a function specified by a table. In such cases, use of some formula expressing integrals in terms of finite differences of the integrand is preferable. It is advisable to use central-difference integration formulae, since in these the coefficients converge to zero. In terms of the central-difference notation for finite differences, the main formulae are:

For a single integration

$$(\delta y)_{\frac{1}{2}} = y_1 - y_0 = \tfrac{1}{2}(\delta x)\left[y_0' + y_1' - \tfrac{1}{12}(\delta^2 y_0' + \delta^2 y_1')\right.$$
$$\left. + \tfrac{11}{720}(\delta^4 y_0' + \delta^4 y_1') - \ldots\right], \tag{9.13}$$

and for a two-fold integration

$$(\delta^2 y)_0 = y_1 - 2y_0 + y_{-1} = (\delta x)^2\left[y_0'' + \tfrac{1}{12}\delta^2 y_0'' - \tfrac{1}{240}\delta^4 y_0'' + \ldots\right]. \tag{9.14}$$

In each of these formulae, terms beyond the first involve the value of the integrand at the end of the interval of integration, but this only comes into the fourth-order term in (9.14), whereas it comes into the main term in (9.13). One result is that direct two-fold integration from y'' to y, without calculating y', is a more satisfactory process, from the point of view of numerical work, than single integration.

In general, if it is required to use these integration formulae, the values of y_1' and the second-difference terms in (9.13), and the second-difference terms in (9.14), have to be obtained by successive approximation, or by making estimates and adjusting them until the estimates agree with the results of carrying out the integration. However, in three cases, namely, (i) linear first-order equations, (ii) n-th order equations linear in the $(n-1)$-th derivatives, and (iii) n-th order equations with the $(n-1)$-th derivative absent, and linear in the $(n-2)$-th derivative, there are available integration formulae which involve no estimation or successive approximation and are correct to a higher order in (δx) than is possible in the general case (53). It is necessary both to have methods suitable for the general case, and to be ready to take advantage of any features of a particular equation which may make possible the use of a better though more special method.

For work with an automatic machine, it seems advisable to use a method which does not involve any estimation, and possible revision of estimates, step by step as the solution proceeds. Then it would appear at first sight that one is restricted to using only the first one or two terms in the integration formulae (9.13) and (9.14), and that it would be necessary to use a small interval of integration to avoid accumulation of appreciable truncation errors. However, this can be avoided by an iterative process, using in the integration formulae in the $(n+1)$-th iteration the differences of the integrand calculated in the n-th iteration. Such a method, in which the contributions forming the higher differences are expressed in terms of the differences of the solution y rather than those of the integrand y' or y'', has recently been studied by Fox and Goodwin (37), with encouraging results. When an automatic machine is used to carry out such a process, a considerable storage capacity is required, since the whole set of function values y_n obtained in one repetition of the iterative process has to be stored for use in the next repetition of the process.

An alternative general method of correcting for the truncation errors is to carry out two independent integrations with different interval lengths δx, and to use the difference between the results to estimate and correct for the truncation error. Care is necessary in this method to ensure that the correction for the truncation error is not vitiated by rounding-off errors.

9.6. Solution of Ordinary Differential Equations With Two-Point Boundary Conditions

The method of evaluating a solution of a differential equation with two-point boundary conditions by evaluating a number of trial solutions is often unsatisfactory, first because the solution is often so sensitive to the trial values of the adjustable initial conditions and parameters, and secondly because of the large number of trial solutions which may have to be evaluated, especially if there is more than one condition to be satisfied. In such a method the solution is approached through a sequence of functions each of which *does* satisfy the equation at all points, but does *not* satisfy all the required conditions. An alternative is the use of an iterative method in which the solution is approached through a sequence of functions each of which *does* satisfy *all* the conditions but does *not* necessarily satisfy the equation.

For example, in the case of the equation

$$y''' = -(1 + y^2)\, y'' \tag{9.15}$$

with conditions

$$y(0) = y'(0) = 0, \qquad y'(\infty) = 1, \tag{9.16}$$

consider the sequence of functions defined by

$$y_{n+1}''' = -(1 + y_n^2)\, y_{n+1}''. \tag{9.17}$$

The solution of (9.17), subject to conditions (9.16) on y_{n+1}, is

$$y_{n+1} = \frac{\int_0 \int_0 \left[\exp\left\{ -\int_0 (1 + y_n^2)\, dx \right\}\right] dx dx}{\int_0^\infty \left[\exp\left\{ -\int_0 (1 + y_n^2)\, dx \right\}\right] dx} \tag{9.18}$$

This may seem a complicated form of the original equation; but it contains the boundary conditions as well, and an iterative process based on this formula converges satisfactorily and is very convenient for numerical work. An advantage of the iterative process in this case is that in each integration the integrand is known over the whole range of x, so that central-difference integration formulae involving high orders of differences can be used, and considerable accuracy attained, without the use of estimation or of very small intervals of integration.

As with the iterative method for ordinary equations with one-point boundary conditions, a considerable storage capacity is required when an automatic calculating machine is used to carry out the numerical work involved in the use of such a method. Even for the simple equation (9.15) a capacity of about 300 numbers would probably be needed, and for more complicated equations, or for a set of simultaneous equations, the capacity required would be several times larger.

For the solution of the set of simultaneous equations (7.1) with the two-point boundary conditions (7.2), such a method has been found very satisfactory for numerical work with a desk machine (24), and it would be admirable for work with an automatic machine of adequate capacity.

Another method of evaluating the solution of equation (9.15) is by minimising the integral

$$\int_0^\infty [y''' + (1 + y^2) \, y'']^2 \, dx$$

subject to the conditions (9.16). Very little work has been done on methods of this kind from the point of view of the application of high-speed automatic machines.

Another kind of problem is one in which one or more of the parameters, which have to be determined in order that the solution should satisfy the specified conditions, are not boundary conditions but constants in the equation itself for which characteristic values are required, as for example in

$$y'' + [f(x) + A] \, y = 0 \qquad (9.19)$$

with $f(x)$ given for $x > 0$, and conditions

$$y(0) = y(\infty) = 0, \qquad \int_0^\infty y^2 dx = 1. \qquad (9.20)$$

It is possible to solve such a characteristic-value problem by evaluating a set of solutions, with trial values of A, satisfying the condition at $x = 0$; but this is not a satisfactory process to mechanise on account of the extreme sensitiveness of y to the value of A in very many cases. Also if A is complex (see, for example, reference 57), the number of trial solutions required may be large.

In the case of the simple equation (9.19), it is in principle possible to determine the characteristic values of A and corresponding solutions by finding the functions y for which $\int_0^\infty y \left[\dfrac{d^2y}{dx^2} + f(x)y \right] dx$ is stationary subject to the conditions (9.20). But a practical numerical process for doing this on a fully automatic machine has not been worked out. Further, such a method may not be applicable to characteristic-value problems of higher order equations, which are certainly possible though not familiar.

A satisfactory iterative method of treating characteristic-value problems is very much wanted and it is required to be applicable not only to the simple case of equation (9.19) with conditions (9.20), but to equations such as those mentioned in §9.4.

9.7. Partial Differential Equations

In the treatment of partial differential equations, by the differential analyser or by some numerical process, different methods are required according as the field of integration is completely closed by a boundary on which boundary conditions are given, or is open (see fig. 67). In equations relating to physical problems, the form of the boundary conditions is often related to the form of the equation, elliptic equations, such as Poisson's equation, usually having boundary conditions specified on one or more closed surfaces (one of which may be the sphere at infinity) bounding the field of integration, whereas in hyperbolic or parabolic equations, such as the wave equation or the equation of heat conduction, the field of integration is open in the direction of one variable (physically, often the time variable). The two forms of boundary condition correspond roughly to the two-point and one-point forms of boundary condition for an ordinary differential equation.

In the application of a digital machine to partial differential equations, it is necessary to replace derivatives with respect to all variables by finite differences (see fig. 67). For an elliptic equation, this gives a number of simultaneous equations enough to determine the solution at all mesh-

OPEN BOUNDARY:

Ex:
$(y = t)$.

$$\frac{\partial \theta}{\partial t} = \frac{\partial^2 \theta}{\partial x^2}$$

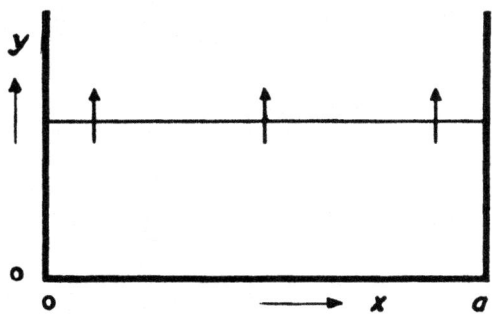

CLOSED BOUNDARY:

Ex:

$$\frac{\partial^2 V}{\partial x^2} + \frac{\partial^2 V}{\partial y^2} = 1$$

Finite difference form:

$$\frac{V_1 + V_2 + V_3 + V_4 - 4V_0}{a^2} = 1$$

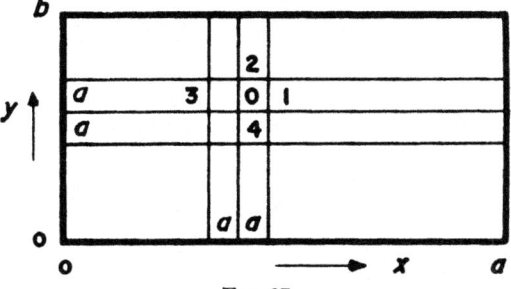

FIG. 67

points of the finite-difference net; the problem is thus reduced to that considered in §9.3. The process of solution may not, however, be quite straightforward. Difficulties have been found in applying Southwell's relaxation technique to the finite-difference form of

$$\frac{\partial^2 V}{\partial x^2} + \frac{\partial^2 V}{\partial y^2} - k^2 V = 0$$

over a region of extent greater than about $2\pi/k$ in either variable, and it seems likely that similar difficulties might occur in other methods of handling the finite-difference equations. The point is one which requires investigation.

For a parabolic or hyperbolic equation, it is not always obvious how best to replace derivatives by finite differences. Consider, for example, the equation of heat conduction in one dimension, in reduced form

$$\frac{\partial \theta}{\partial t} = \frac{\partial^2 \theta}{\partial x^2} \tag{9.21}$$

with given boundary conditions

$$\begin{aligned} \theta &= f_1(t) \text{ at } x = 0 \\ \theta &= f_2(t) \text{ at } x = a \end{aligned} \Bigg\} \, t > 0$$

$$\theta = F(z) \text{ at } t = 0, \quad 0 \leqslant x \leqslant a \tag{9.22}$$

If δx, δt are the finite intervals into which the ranges of x and t are divided, let $\theta_m(n)$ stand for the temperature at $x = m\delta x$, $t = n\delta t$. Finite difference approximations to $\partial \theta/\partial t$ and $\partial^2 \theta/\partial x^2$ at (m,n) are

$$(\partial \theta/\partial t)_{m,\,n} = [\theta_m(n+1) - \theta_m(n-1)]/2\delta t, \tag{9.23}$$

$$(\partial^2 \theta/\partial x^2)_{m,\,n} = [\theta_{m+1}(n) - 2\theta_m(n) + \theta_{m-1}(n)]/(\delta x)^2. \tag{9.24}$$

Equating these two gives

$$\theta_m(n+1) = \theta_m(n-1) + \frac{2\delta t}{(\delta x)^2} [\theta_{m+1}(n) - 2\theta_m(n) + \theta_{m-1}(n)]. \tag{9.25}$$

This formula looks very attractive for numerical work, since if θ has been evaluated up to time $t = n\delta t$, each single value of θ at time $(n+1)\delta t$ such as that represented by a circle in fig. 68a, is expressed in terms of values of θ at previous times, such as those represented by crosses in fig. 68a, all of which are known. But it is quite unusable as a formula for a practical method, since effects of rounding-off errors build up rapidly and in a quite uncontrollable way; this situation cannot be remedied by taking smaller intervals (δt) (see reference 28).

On the other hand, if the finite-difference approximation to $(\partial\theta/\partial t)_{m,\,n+\frac{1}{2}}$ namely

$$(\partial\theta/\partial t)_{m,\,n+\frac{1}{2}} = [\theta_m(n+1) - \theta_m(n)]/\delta t$$

is equated to the result of substituting the approximation (9.24) in

$$(\partial^2\theta/\partial x^2)_{m,\,n+\frac{1}{2}} = \tfrac{1}{2}\left[(\partial^2\theta/\partial x^2)_{m,\,n+1} + (\partial^2\theta/\partial x^2)_{m,\,n}\right],$$

the process of solution of the resulting system of equations is stable, though each gives only a relation between values of θ at three successive points $(m-1)$, m, $(m+1)$ at time $(n+1)\delta t$ (see fig. 68b). This method has been used successfully for the calculation of heat flow in a substance in which a chemical change is occurring, with evolution of heat, at a rate depending on the local temperature, so that there is at each point a chemical kinetic equation as well as the equation of heat conduction (28).

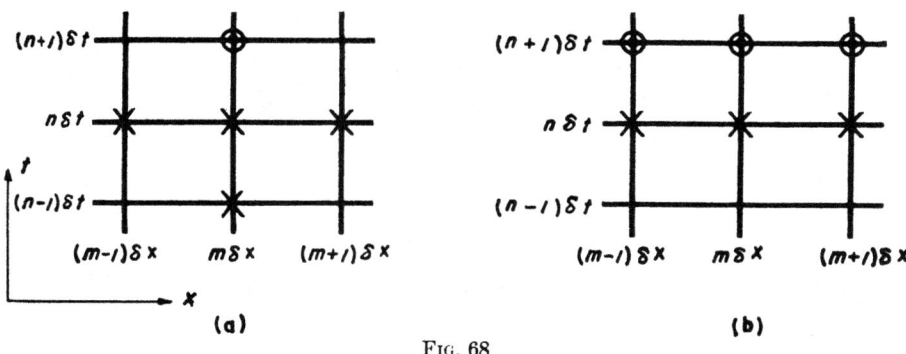

FIG. 68

Another possible method of handling a partial differential equation is to convert it into an integral equation and solve this by an iterative process. This may involve the numerical evaluation of a large number of multiple integrals, and would be almost impracticable without a means of carrying out a really large number of multiplications, but might be practicable with machines of the capacity and speed now contemplated.

For hyperbolic equations, it is probable that some finite-difference form of the method of characteristics will be the most effective.

In many physical contexts, a partial differential equation for the space variation of some quantity in a steady state may be regarded as a special case of a more general equation for a non-steady state. For example

$$\frac{\partial^2\theta}{\partial x^2} + \frac{\partial^2\theta}{\partial y^2} = -1 \qquad (9.26)$$

can be regarded as a special case of the equation of heat conduction with constant and uniform rate of generation of heat

$$\frac{\partial \theta}{\partial t} = \frac{\partial^2 \theta}{\partial x^2} + \frac{\partial^2 \theta}{\partial y^2} + 1, \qquad (9.27)$$

and it may be practicable to evaluate the solution of an equation such as (9.26) as the asymptotic solution of (9.27) as t tends to infinity. If equation (9.26) is replaced by a set of finite-difference equations, the residuals of these equations correspond to the values of $\partial \theta / \partial t$ in equation (9.27); the time which elapses in the process of minimising the sum of the squares of these residuals can be regarded as corresponding roughly to the additional time variable introduced in equation (9.27).

REFERENCES

1. Adcock, W. A. Rev. Sci. Insts. **19**. 181, 1948.
2. Aiken, H. H., and Hopper, G. H. Elect. Eng. **65**, 384, 449, 552, 1946.
3. Aitken, A. C. Proc. Roy. Soc. Edin. **46**, 289, 1926 (§8).
4. Aitken, A. C. Proc. Roy. Soc. Edin. **57**, 269, 1937 (§8).
5. Amble, O. Journ. Sci. Insts. **23**, 284, 1946.
6. Babbage, C. "Passages from the Life of a Philosopher" (Longmans, London, 1864).
7. Babbage, H. P. "Babbage's Calculating Engines" (London, Spon Ltd., 1889).
8. Beard, J. E. Journ. Inst. of Actuaries **71**, 193, 1941.
9. Berry, C. E., Wilcox, D. E., Rock, S. M., and Washburn, H. W. J. Appl. Phys. **17**, 262, 1946.
10. Beuken, L. Econom. Tech. Tijdschrift **19**, 43, 1939.
11. Blackett, P. M. S., and Williams, F. C. Proc. Camb. Phil. Soc. **35**, 494, 1939.
12. Booth, A. D. Journ. Chem. Phys. **15**, 415, 1947.
12a. Booth, A. D. Ref. 96a, p. 286.
13. Born, M., Fürth, R., and Pringle, R. W. Nature **156**, 756, 1945.
14. Bratt, J. B., Lennard-Jones, J. E., and Wilkes, M. V. Proc. Camb. Phil. Soc. **35**, 485, 1939.
15. British Association Mathematical Tables, Vol. II (London, 1932) ; Introduction.
16. Burks, A. W. Proc. Inst. Radio Eng. **35**, 756, 1947.
17. Bush, V. Journ. Franklin Inst. **212**, 447, 1931.
18. Bush, V., and Caldwell, S. H. Journ. Franklin Inst. **240**, 255, 1945.
19. Callender, A., Hartree, D. R., and Porter, A. Phil. Trans. Roy. Soc. **235**, 415, 1936.
20. Callender, A., Hartree, D. R., Stevenson, A. B., and Porter, A. Proc. Roy. Soc. **161**, 460, 1937.
20a. Campbell, R. V. D. Ref. 63, p. 69.
21. Cesareo, O. Bell Laboratories Record **24**, 457, 1946.
22. Comrie, L. J. Journ. Sci. Insts. **21**, 129, 1944.
23. Comrie, L. J. Nature **158**, 567, 1946.
24. Cope, W. F. and Hartree, D. R. Phil. Trans. Roy. Soc. **241**, 1, 1948.
25. Copple, C., Hartree, D. R., and others. Journ. Inst. El. Eng. **85**, 56, 1939.
26. Couffignal, L. "Les grandes machines mathematiques" (Edition de la Revue d'optique, Paris, 1948), Part II.
27. Crank, J. "The Differential Analyser" (Longmans, Green and Co., London, 1948).
28. Crank, J., and Nicolson, P. Proc. Camb. Phil. Soc. **43**, 50, 1947.
29. Dietzold, R. L. Bell Laboratories Record **16**, 130, 1937.
30. Domb, C. Proc. Camb. Phil. Soc. **45**, 237, 1949.
31. Encyclopaedia Britannica, 14th ed., article on "Mathematical Instruments."
32. Eyres, N. R., Hartree, D. R., and others. Phil. Trans. Roy. Soc. **240**, 1, 1946.
33. Fairweather, A., and Ingham, J. Journ. Inst. Elect. Eng. **88**, 330, 1941.
34. Forrester, J. W. Ref. 63, p. 125.
35. Fox, L. Proc. Roy. Soc. **190**, 31, 1947.
36. Fox, L. Quart. Journ. Mechanics and Appl. Math. **1**, 253, 1948.
37. Fox, L., and Goodwin, E. F. Proc. Camb. Phil. Soc. **45**, 373, 1949.
38. Fox, L., Huskey, H. D., and Wilkinson, J. H. Quart. Journ. Mech. and Appl. Math. **1**, 149, 1948.

39. Goldstine, H. H. and A. Math. Tables and Aids to Computation **2**, 97, 1946.

40. Goldstine, H. H. and von Neumann, J. Planning and Coding of Problems for an Electronic Computing Instrument (Institute for Advanced Study, Princeton, 1947; duplicated).

41. Goodlet, B., Edwards, F. S., and Perry, F. R. Journ. Inst. Elect. Eng. **69**, 695, 1930.

42. Goldberg, E. A., and Brown, G. W. Journ. Appl. Phys. **19**, 339, 1948.

43. Gray, T. S. Journ. Franklin Inst. **212**, 77, 1931.

44. Haegg, G., and Laurent, T. Journ. Sci. Insts. **23**, 155, 1946.

45. Hartree, D. R. Phys. Rev. **46**, 738, 1934.

46. Hartree, D. R. Proc. Camb. Phil. Soc. **33**, 223, 1937.

47. Hartree, D. R. Mathematical Gazette **22**, 343, 1938.

48. Hartree, D. R. Aeronautics Research Committee, Fluid Motion Panel Paper 417, 1939. (Reports and Memoranda of the A.R.C., No. 2427).

49. Hartree, D. R. Journ. Inst. Electrical Engineers **90**, 435, 1943.

50. Hartree, D. R. Nature **158**, 500, 1946.

51. Hartree, D. R. The Times (London), Nov. 7, 1946.

52. Hartree, D. R. Journ. Sci. Insts. **24**, 172, 1947.

53. Hartree, D. R. Ref. 85; Lecture 5, p. 21.

54. Hartree, D. R. Eureka, March 1948, p. 13.

55. Hartree, D. R. Proc. Camb. Phil. Soc. **45**, 230, 1949.

56. Hartree, D. R., and Ingham, J. Mem. and Proc. Manchester Lit. Phil. Soc. **83**, 1, 1938.

57. Hartree, D. R., Michel, J. G. L., and Nicolson, P. Meteorological factors in radio propagation (Joint report of Physical and Royal Meteorological Societies, London 1947); p. 127.

58. Hartree, D. R., and Nuttall, A. K. Journ. Inst. Electrical Engineers **83**, 643, 1938.

59. Hartree, D. R., and Porter, A. Mem. and Proc. Manchester Lib. Phil. Soc. **79**, 51, 1935.

60. Hartree, D. R., and Porter, A. Journ. Inst. Elect. Eng. **83**, 648, 1938.

61. Hartree, D. R., and Womersley, J. R. Proc. Roy. Soc. **161**, 353, 1937.

62. Harvard Computation Laboratory Annals, Vol. I (Harvard University Press, 1946).

63. Harvard Computation Laboratory Annals, Vol. XVI, 1948 (Proceedings of a Symposium on large scale digital calculating machinery).

64. Hazen, H. L., and Brown, G. S. Journ. Franklin Inst. **230**, 19 and 183, 1940.

65. Hazen, H. L., Jaeger, S. S., and Brown, G. S. Rev. Sci. Insts. **7**, 353, 1936.

66. Hazen, H. L., Schurig, O. R., and Gardner, M. E. Trans. Amer. Inst. Elect. Eng. **49**, 1102, 1930.

67. Horsburgh, E. M. (Editor) Handbook of the Napier Tencentenary Exhibition (Royal Society of Edinburgh, 1914); reprinted under title "Modern Instruments of Calculation" (Bell and Co., London, 1914).

68. Hotelling, H. Ann. Math. Statistics **14**, 1, 1943.

69. Householder, A. S., and Landahl, H. D. "Mathematical Biophysics of the Central Nervous System" (Principia Press, 1945) Part III.

70 International Business Machines Corporation, Booklet on the I.B.M. Selective Sequence Electronic Calculator (New York, 1948).

71 Jackson, R., Sarjant, R. J., and others. Journ. Iron and Steel Inst. **150**, 211, 1944.

Kelvin, Lord (*See* Thomson, W.)

72. Kornei, O. Ref. 63, p. 223.

73. Kuehni, H. P., and Lorraine, R. S. Trans. Amer. Inst. Elect. Eng. **57, 67,** 1938.

74. Lamb, H. Hydrodynamics (Cambridge 3rd ed., 1906) §278.

75. Lemaitre, G., and Vallarta, M. S. Phys. Res. **49, 719,** 1936.

76. Macewen, D., and Reevers, C. A. Journ. Sci. Insts. **19, 150,** 1942.

77. Mallock, R. C. C. M. Proc. Roy. Soc. **140, 457,** 1933.

78. Manning, M. E., and Millman, S. Phys. Rev. **49, 848,** 1936.

79. Massey, H. S. W., Buckingham, R. A., Whlie, J., and Sullivan, R. Proc. Royal Irish Acad. **45,** 1, 1938.

80. McCulloch, W. S., and Pitts, W. Bull. Math. Biophysics **5, 115,** 1943.

81. Menebrea, L. F. Bibliothèque Universelle de Genève, No. **82** (1842).

82. Menebrea, L. F. Scientific Memoirs (London, ed. R. Taylor) **3, 666,** 1842.

83. Mercner, R. O. Bell Laboratories Record **16, 135,** 1937.

84. Michel, J. G. L. Journ. Sci. Insts. (in press).

85. Moore School of Electrical Engineering, Univ. of Pennsylvania "Theory and Techniques for the Design of Electronic Digital Computors". (1947-8, duplicated).

86. Murray, F. J. "Theory of Mathematical Machines" (Kings Crown Press, New York, 1947).

86a. Myers, D. M. Journ. Sci. Insts. **16, 209,** 1939.

87. Nature **146,** 319, 1940.

88. von Neumann, J., and Goldstine, H. H. Bull. Amer. Math. Soc. **52,** 1021, 1947.

89. Parker, W. W. Trans. Amer. Inst. Elect. Eng. **60,** 977, 1941.

90. Paschkis, V., and Baker, H. D. Heat Treatment and Forging **27,** 375, 1941.

91. Paschkis, V., and Baker, H. D. Trans. Amer. Soc. Mech. Eng. **64,** 102, 1942.

92. Pepinsky, R. J. Appl. Phys. **18,** 601, 1947.

93. Rajchman, J. Ref. 63, p. 163.

94. Reeves Instrument Corporation (New York), The Reeves Analogue Computor (REAC).

95. Richardson, L. F. Phil. Trans. Roy. Soc. **226,** 299, 1927.

96. Riemann, B. Gött. Abh. **8,** 43, 1858 and Gesammelte Werke, 2te. Aufl. (Teubner, Leipzig, 1892) p. 157.

96a. Royal Society. Discussion on computing machines. Proc. Roy. Soc. **195,** 265, 1948.

97. Samuelson, P. A. Journ. Math. and Phys. **24,** 131, 1945.

98. Shannon, C. E. Trans. Amer. Inst. Elect. Eng. **57,** 713, 1938.

99. Sharpless, T. K. Electronics **20** (11), 134, 1947.

100. Sharpless, T. K. Ref. 63, p. 63.

101. Southwell, R. V. "Relaxation Methods in Engineering Science" (Oxford, 1940).

102. Southwell, R. V. "Relaxation Methods in Theoretical Physics" (Oxford, 1946).

102a. Tabor, L. B. Ref. 63, p. 31.

103. Taylor, R., and Thomas, G. B. (in press).

104. Thomson, J. Proc. Roy. Soc. **24,** 262, 1876, and Thomson, W., and Tait, P. G. Treatise on Natural Philosophy, Vol. I, Appendix B', III.

105. Thomson, W. (Lord Kelvin). Proc. Roy. Soc. **24, 266,** 1876, and Thomson, W., and Tait, P. G. Treatise on Natural Philosophy, Vol. I, Appendix B', IV.

106. Thomson, W. (Lord Kelvin). Proc. Roy. Soc. **24, 269,** 1876, and Thomson, W., and Tait, P. G. Treatise on Natural Philosophy, Vol. I, Appendix B', V.

107. Thomson, W. (Lord Kelvin). Proc. Roy. Soc. **24, 271,** 1876, and Thomson, W., and Tait, P. G. Treatise on Natural Philosophy, Vol. I, Appendix B', VI.

108. Thomson, W. (Lord Kelvin). Proc. Roy. Soc. **27**, 371, 1878, and Thomson, W., and Tait, P. G. Treatise on Natural Philosophy, Vol. I, App. B′, VII.

109. Travis, I., and Weygandt, E. V. Trans. Amer. Inst. Elect. Eng. **57**, 423, 1938.

110. Tweedie, C. Reference 67, section G, I, on "Integraphs."

111. Weekes, K., and Wilkes, M. V. Proc. Roy. Soc. **192**, 80, 1947.

112. Wilkes, M. V. Electronic Eng. **19**, 104, 1947.

113. Wilkes, M. V., and Renwick, W. Electronic Eng. **20**, 208, 1948.

114. Wilkes, M. V. Ref. 96a, p. 275.

115. Wilkinson, J. H. Ref. 96a, p. 285.

116. Williams, F. C. Ref. 96a, p. 279.

117. Williams, F. C., and Kilburn, T. Nature **162**, 487, 1948.

118. Williams, F. C., and Kilburn, T. Journ. Inst. Elect. Eng. **96**, 81, 1949.

119. Williams, F. C., and Uttley, A. M. Journ. Inst. Elect. Eng. Vol. 93, Pt. IIIa (Proc. Radiolocation Convention) pp. 317 and 1256, 1948.

120. Williams, S. B. Bell Laboratories Record **25**, 49, 1947.

121. Williams, S. B. Ref. 63, p. 41.

122. Wilson, A. H. Proc. Camb. Phil. Soc. **36**, 365, 1938.

Additional References

123. Bloch, R. M., Campbell, R. V. D., and Ellis, M. Math. Tables and Aids to Computation, 3, 286, 317, 1948.

124. Kilburn, T. Nature, **164**, 684, 1949.

125. Turing, A. M. Proc. Lond. Math. Soc. **42**, 230, 1936.

126. Wilkes, M. V. Nature, **164**, 341, 557, 1949.

127. Wilkes, M. V. Journ. Sci. Insts. **26**, 217, 1949.

128. Wilkes, M. V. Journ. Sci. Insts. **26**, 385, 1949.

NAME INDEX

135

SUBJECT INDEX